The Impact of Civilisation on the Biology of Man

Papers from a symposium held on 11-12
September 1968 at Canberra, sponsored by the
Australian Academy of Science

The Impact of Civilisation on the Biology of Man

Edited by **S. V. Boyden**

Australian National University Press — Canberra 1970

Registered at the G.P.O. for transmission by post as a book

Library of Congress Catalog Card no. 76-115 889

National Library of Australia Card no. and
ISBN 0 7081 0080 5

Contributors

Dr H. F. Barnes, Urban Biology Group, John Curtin School of Medical Research, A.N.U.

Dr Barnes received her original training in anthropology and medicine at Cambridge University and she has lived in Africa and Norway. She is interested in child development, small group dynamics and in 'what makes humans human'.

Dr S. V. Boyden, F.A.A., Urban Biology Group, John Curtin School of Medical Research, A.N.U.

Dr Boyden became a member of the Royal College of Veterinary Surgeons in 1947, and Ph.D. (Cambridge) in 1951. He has carried out research in immunology at the Rockefeller Institute (New York), the Pasteur Institute (Paris), and from 1952 to 1960 was Chief of the Tuberculosis Immunization Research Centre (WHO) in Copenhagen. He continued work in this general field in the Department of Experimental Pathology at the John Curtin School of Medical Research from 1960 to 1966, when he gave up immunology in order to study the biology of civilisation. At present he is Head of the Urban Biology Group at the John Curtin School of Medical Research, A.N.U.

Sir Macfarlane Burnet, O.M., F.R.S., F.A.A.

Sir Macfarlane Burnet was from 1944 to 1965 Director of the Walter & Eliza Hall Institute, Melbourne and Professor of Experimental Medicine at Melbourne University. He shared the Nobel Prize for Physiology and Medicine in 1960, and is the author of many publications in the fields of virology and immunology. He was President of the Australian Academy of Science from 1965 to 1969. His interests extend beyond the range of his specialisation to include the social implications of science and human biology.

Dr F. W. Clements, School of Public Health and Tropical Medicine, University of Sydney, N.S.W.

Dr Clements is currently Senior Lecturer in Child Health and Nutrition, University of Sydney. From 1949 to 1951 he was Chief, Nutrition Section (WHO), Geneva. For eleven years prior to 1949 he was Director, Institute of Anatomy, Canberra, and Head of Nutrition Unit, Commonwealth Department of Health. He has carried out studies and surveys into various aspects of child health and nutrition for the World Health Organization in Ceylon, Formosa, Philippines, India, and Egypt. He is the author of several books on nutrition.

Professor R. J. Dubos, The Rockefeller University, New York.

Professor Dubos was born and educated in France and moved to the United States of America in 1924. He joined the Rockefeller Institute (now Rockefeller University) in 1927, where he carried out research in bacteriology, which led amongst other things to the first crystallisation of an antibiotic. His bacteriological work culminated in a classic book *The Bacterial Cell.* Subsequently work on tuberculosis led to an interest in the host and his environment and so to Dubos's present concern with man and the environmental situation, portrayed in such books as *Man Adapting* and *So Human an Animal.*

Professor F. Fenner, F.R.S., F.A.A., John Curtin School of Medical Research, A.N.U.

Professor Fenner has been Director of the John Curtin School of Medical Research since 1967. He graduated in medicine at the University of Adelaide in 1938, served in the AIF from 1940 to 1946, worked for two and a half years at the Walter & Eliza Hall Institute in Melbourne, and was appointed Professor of Microbiology in the J.C.S.M.R. in 1949. He is the author of *The Production of Antibodies* (with F. M. Burnet, 1949); *Myxomatosis* (with F. N. Ratcliffe, 1965); *The Biology of Animal Viruses* (1968) and numerous papers on virology and epidemiology.

Dr S. B. Furnass, University Health Service, A.N.U.

Dr Furnass was appointed as Director of the Australian National University Health Service in 1966. A graduate of the University of Oxford, he was previously research assistant in the Department of Medicine at the Middlesex Hospital, London, and consultant in private practice in Canberra.

Dr G. McBride, Animal Behaviour Unit, University of Queensland, St Lucia, Queensland.

At present Dr McBride is Senior Lecturer in Animal Behaviour at the University of Queensland and he is engaged in research in various aspects of animal social organisation and communication. His major publications are 'A General Theory of Social Organization and Behaviour' (Veterinary Science Papers, 1, 1964); *Society Evolution* (1966); 'Behavioural Measures of Social Stress' (in *Adaptation in Domestic Animals,* 1968, ed. E.S.E. Hafez).

Dr C. A. C. Mims, Department of Microbiology, John Curtin School of Medical Research, A.N.U.

Dr Mims read zoology at University College, London, from 1943 to 1947 and then studied medicine at Middlesex Hospital Medical School, London, where he was Demonstrator in the Department of

Physiology from 1947 to 1949, graduating as M.B.B.S. in 1952. He spent from 1953 to 1956 at the East African Virus Research Institute, Entebbe, Uganda, working on virus infections of mosquitoes, laboratory mice, various wild animals, and man. In 1957 he came to Canberra to join the Department of Microbiology, A.N.U. Since then he has been working on the pathogenesis of virus infections, especially virus infections of the liver, brain, skin, and foetus. He has developed an interest in the role of stress in susceptibility to virus diseases.

Dr J. M. Rendel, F.A.A., Division of Animal Genetics, CSIRO, Epping, N.S.W.

Dr Rendel joined the Agricultural Research Council after the war in 1946, and in 1951 he joined CSIRO to lead a small section housed in Sydney University to study quantitative genetics and animal breeding in Australian livestock and start courses in the University. He was appointed Chief of the Division of Animal Genetics when the Division was formed in 1959. His main academic interest has been in quantitative genetics and the genetic control of development variability.

Discussants

Dr R. E. Barwick, Department of Zoology, School of General Studies, A.N.U.

Professor J. H. Bennett, R. A. Fischer Laboratories, Department of Genetics, University of Adelaide, S.A.

Professor W. D. Borrie, Research School of Social Sciences, A.N.U.

Mr A. Brownlea, School of Earth Sciences, Macquarie University, North Ryde, N.S.W.

Professor D. G. Catcheside, F.R.S., F.A.A., Research School of Biological Sciences, A.N.U.

Dr J. D. Freeman, Department of Social Anthropology, Research School of Pacific Studies, A.N.U.

Dr E. Hackett, The Institute of Medical and Veterinary Science, Adelaide, S.A.

Professor B. S. Hetzel, Department of Social and Preventive Medicine, Monash Medical School, Alfred Hospital, Prahran, Victoria.

Dr L. Hiatt, Department of Anthropology, University of Sydney, N.S.W.

Dr E. H. Hipsley, Department of Health, Australian Institute of Anatomy, Canberra.

Professor M. R. Lemberg, F.R.S., F.A.A., Institute of Medical Research, The Royal North Shore Hospital, St Leonards, N.S.W.

Dr M. McCall, Department of Medicine, University of Western Australia, Crawley, W.A.

Mr K. Myers, Division of Wildlife Research, CSIRO, Canberra.

Dr P. Nestel, Department of Clinical Science, Canberra Community Hospital, Canberra.

Mr N. Oram, Department of Anthropology, Research School of Pacific Studies, A.N.U.

Professor K. A. Provins, Department of Psychology, School of General Studies, A.N.U.

Professor S. D. Rubbo, Department of Microbiology, University of Melbourne, Victoria.*

Dr M. Simons, Department of Surgical Research, Royal Children's Hospital Research Foundation, Parkville, Victoria.

Dr R. Shatin, International Society for Research into Nutrition and Civilisation Diseases, Alfred Hospital, Prahran, Victoria.

Professor N. F. Stanley, Department of Microbiology, University of Western Australia, School of Medicine, Perth, W.A.

Dr R. Story, Division of Land Research, CSIRO, Canberra.

Mr M. V. Tracey, Division of Food Preservation, CSIRO, Ryde, N.S.W.

Mr Gavin Walkley, Australian Institute of Urban Studies, Adelaide, S.A.

Dr R. H. C. Wells, National Health and Medical Research Council, Commonwealth Department of Health, Canberra.

* Deceased.

Contents

Tables

Figures

Introduction

Ten thousand or so years ago, a development of outstanding biological significance took place. Some of our ancestors, for the first time in the history of mankind, gave up the hunter-gatherer way of life for a more settled existence based on agriculture and the breeding of domestic animals. With this development human culture acquired a new significance in the biosphere, introducing new forces which were in many ways quite unlike those which had controlled the ecology and evolution of species during the preceding 2,000 to 3,000 million years of life on earth. Since that time these new forces of culture have been interacting on an ever-increasing scale with the ancient forces of nature, and quite recently this interaction has intensified to an extent which many scientists believe constitutes a very grave threat to the survival, let alone the well-being, of humanity.

The processes of civilisation have had, apart from their manifold and complex effects on the biosphere as a whole, a profound influence on the biology of man himself. Thus, the human organism in modern Western society is exposed to biological conditions which differ in many important ways from those of the pre-neolithic era. These new conditions differ, therefore, from those which, through natural selection, determined the biological characteristics of the human species.

How the human organism is reacting biologically to these changes in its conditions of life is of crucial significance in relation to our attempts to understand the contemporary human situation in scientific terms. It is, therefore, remarkable that this problem as a whole has received so little attention from biological or social scientists. The purpose of this symposium was to stimulate discussion, debate and exchange of ideas on this extremely important but at present neglected aspect of human biology.

Needless to say, in a two-day symposium it was possible to deal with only certain aspects of this broad subject, and the topics for discussion were selected on the following basis. Firstly, since the theme of the meeting was the impact of civilisation on the biology of man, it was considered appropriate to include a paper, for purposes of comparison and contrast, on the biology of *Homo sapiens* before the advent of civilisation, 'The Biology of Pre-Neolithic Man' by H. F. Barnes.

When the conditions of life of an animal population deviate from those to which it has become, through natural selection, genetically adapted, some signs of biological maladjustment are almost inevitable.

Since the human species is no exception to this principle, two papers were included on aspects of biological maladjustment in man resulting from environmental changes accompanying the development of civilisation ('The Effects of Changing Social Organisation on Infectious Diseases' by F. Fenner and 'Changes in Non-Infectious Diseases Associated with the Processes of Civilisation' by S. B. Furnass).

The effect of the new conditions associated with civilisation on the genetic constitution of human populations is a subject that has already received a great deal of attention from eugenicists, and many authors have expressed concern about the dysgenic effects of modern medicine. While there is general agreement that social developments since the neolithic revolution have resulted in changes in selection pressures and in gene frequencies in human populations, the crucial problem of the *rate* at which such changes may be taking place is still controversial. This question is dealt with by J. Rendel in his paper 'The Time Scale of Genetic Change'.

As Dr Rendel points out, the rate of genetic change in human populations in response to the new conditions of civilisation is very slow, and for this and other reasons it is clear that man's more or less successful adaptation to the recent environmental changes to which he has been exposed has been due not so much to genetic or evolutionary adaptation as to the processes of 'cultural adaptation'. In view of the outstanding importance of cultural adaptation in terms of the survival and well-being of man in the modern world, and because there has been surprisingly little discussion on the topic up to now, a paper was included on 'Cultural Adaptation to Biological Maladjustment' by S. Boyden.

No attempt was made at the meeting to compile a list of the biologically significant changes that have occurred in the conditions of life of the human organism as a consequence of the processes of civilisation. Although well documented information of this sort would indeed be very useful, a symposium is not the appropriate place for its presentation. However, three papers were presented dealing with certain aspects of environmental change and their biological implications. In the first of these, 'Some Effects of Different Diets', F. Clements discussed the modifications which have occurred in human diets since the neolithic development, and particularly during the past hundred years or so. Two less easily definable, more controversial, but nonetheless very important aspects of environmental change were considered in the papers 'Social Adaptation to Crowding' by G. McBride and 'Stress in Relation to the Processes of Civilisation' by C. Mims.

In the final paper, R. Dubos deals in a general way with problems of adaptation in man, and he emphasises particularly the great importance of early biological experiences of the individual in relation to his later capacity to cope with various kinds of environmental challenge.

The discussion after each contribution was opened by invited speakers who had read the papers beforehand. These prepared comments are included in this volume together with part of the open discussion which followed them.

Canberra 1969 S.V.B.

F. M. Burnet

Chairman's opening remarks

Human Biology as the Study of Human Differences

At the present time an International Association of Human Biologists is being initiated. The interim Council of this organisation has defined the scope of human biology as:

'the study of the nature, development, causes and origins of variation in human populations, at the molecular, cellular, tissue, and whole body levels. In particular, emphasis is given to the relationship between genetic and environmental factors in producing this variation, as it occurs both within and between populations.'

The subject does not cover normal anatomy, physiology, and biochemistry in so far as these disciplines are equally applicable to other species of mammal. It concerns such matters as:

At the molecular level—the haemoglobin variants, in particular Hb.S, the disease it is associated with (sickle cell anaemia) and its evolution.

At the cellular level—the immunological grouping of blood cells and leucocytes and their bearing on haemolytic (Rh) disease of the newborn, on the evolution and movement of human races or on forensic matters like disputed paternity.

At the whole body level—skin colour and its anomalies, albinism and freckling, for example; somatotypes, i.e. the classification of body forms.

These are simply examples. At each level there are hundreds of other topics that are relevant to this approach.

Logically, human biology must go further than that, just as soon as the appropriate methodology is available: to the study of human temperament and human behaviour in so far as these can be shown to have a basis, genetic or environmental, that is independent of the immediate circumstances.

Human biology is essentially concerned with the genetic differences between human beings, how these differences have originated and been distributed, and the part played by environment both as a selective agent or as an actively modifying one in making those differences evident. If we are to discuss the impact of civilisation on human biology we must first have an understanding of the genetic structure of the human species, and of the degree to which genetic characteristics, either of the species as a whole or of special groups of men, are adaptations to environmental conditions. As in every species of mammal, an important part of the environment is provided by other members of the same species. In man, probably from the earliest hominids onward, this has progressively become by far the most important element. For most of us the whole environment is man-made and man-controlled—other aspects become important only in natural catastrophes or from breakdown in human control.

We have evolved from near-human primate to human form over about one to two million years and there is no significant skeletal difference between Cro-Magnon man of about 30,000 years ago and many modern west Europeans. The oldest urban civilisations developed in the Middle East about 5,000 years ago. This is a very short period of about 200 human generations, far too short for major mutational changes to influence genetic constitution. When we consider, however, the possibilities of redistribution of genetic qualities already developed in one or other component of the human gene pool during the previous million years, selection and genetic drift between them could well have produced major genetic changes in a human population over that 5,000 years. It has been estimated that on the average, each human being carries eight or ten recessive genes which in double dose would give rise to medically recognisable anomaly or disease. Some are quite disconcertingly common. The gene responsible for cystic fibrosis is present in from 2-8 per cent of people in the communities that have been studied. Amongst the Melanesians of the Trobriand Islands, the gene for albinism is present in around 6 per cent of individuals. It is equally certain that each of us has large numbers of genes which, appropriately paired, would have equally significant effects on temperament, intelligence, and the capacity to develop special skills. Despite the shortness of the history of civilisation, there has been ample scope for changes in the genetic constitution of human populations. Undoubtedly such changes will continue in the future but their magnitude and speed may be inhibited by two new factors specially characteristic of the twentieth century. By the virtual elimination of lethal infectious disease, and hence of early death, we have reduced the intensity of selective processes and by the rapid spread of effective contraception we are also reducing the amount of material available for the winnowing process of selection. Both factors will probably act to slow the process of change. Others, which I shall mention later, may intensify it.

It has been usual to lay stress on the dysgenic effects of medical care as a potent cause for genetic deterioration in the future. It is probably true, as Müller claimed, that if all haemophiliacs could be treated effectively enough to give them the same fertility as normal individuals, the condition would spread through the species like a slow infectious disease until 50 per cent of the population carried the gene. He calculated that this would require some 30,000 generations, a million years, so that the threat is hardly an urgent one. Nevertheless there is undoubtedly a long-term danger. If the standard social pattern of today in Western communities continues indefinitely, there can be no escape from serious genetic deterioration. With essentially random mating as far as medically significant genes are concerned, with a steady slow appearance of new mutations, with small two to four child families and with the survival of at least 90 per cent to reproductive age, there is no scope whatever for natural selection by elimination of the unfit in the Darwinian sense. However, there is nothing to suggest that there will be any positive selective value of medically harmful mutant genes and unless there is, it will be many generations before even relatively common mutations, such as that responsible for haemophilia, will have any effect visible in demographic statistics. There are probably a thousand years ahead of us which we can use to work out the laws of human inheritance and to develop the necessary social conscience to allow a scientific and acceptable application of eugenics to counter the trend. The titles of contributions to this symposium and the current approach to social problems suggest that we are all thinking essentially of what is happening now or what can be expected to happen in, at most, the next hundred years.

There is no reason why we should not take a longer view, and I shall be most interested to hear what Dr Rendel has to say in his talk. My present thought, however, is in regard to such matters as, first, differential fertility, particularly as modified by the acceptability and use of contraceptive techniques, and, second, possible modification of mating groups to allow, for example, the appearance of pockets of population of exceptionally high intelligence in certain academic and science based industrial environments. For the most part, however, we are concerned in this symposium with the non-genetic effects on the human organism of changing present day environment. I shall take the point of view that the average genetic constitution of present day persons is not greatly different from what it was a hundred thousand years ago, well before the advent of any form of pastoral or agricultural activity. Over the historical period—the period of urbanisation—I can conceive a significant effect of the high mortalities in infancy and childhood in providing a genetically based increased resistance to infectious disease, but I cannot see any reason for any other character to be significantly affected. I have seen it suggested that the tendency for men of unusual intellectual power to become celibate members of religious communities during the Middle Ages may have lowered the average intelligence of Western Europeans, but the effect was probably quite insignificant. As far as general bodily health and patterns of behaviour are concerned, one can feel reasonably confident that the genetic aspects of these qualities have not changed significantly since the days of Cro-Magnon man.

In my Boyer series of A.B.C. lectures, given in 1966, I said that 'the problem of today is how to use the intelligence of a relatively small number of men and women to devise ways by which patterns of behaviour laid down in a million years can be modified, tricked and twisted if necessary, to allow a tolerable human existence in a crowded world'. I was pleased when Dennis Gabor, in a contribution to a Royal Society symposium on a related topic, quoted that sentence as expressing the gist of the problem.

Most of the contributions to the present symposium have little direct concern with human behaviour, for the obvious reason that little is known about significant aspects of human behaviour which is expressible in quantitative terms or allows meaningful prediction of future behaviour by individuals or communities. There is, in addition, the widespread implicit dogma that human behaviour has no genetic component and that all anomalies of behaviour can be righted by environmental action, i.e. by education. This appears to be the background of most writing on education and cultural anthropology as well as of classical Marxist philosophy. I shall take the point of view expressed by Washburn and many others (self-evident to anyone who has been a member of a large family) that behaviour has a very important genetic component. All the details of behaviour are learnt, are culturally determined, but the capacity to learn some things more easily than others is genetic, is subject to evolutionary processes, and is all-important in regard to such significant aspects of behaviour as intellectual achievement and level of aggressiveness.

It is probably far too early to forecast what is happening in regard to the distribution in civilised communities of the two genetic qualities of intellect and aggressiveness. One can contrast the qualities of, on the one hand, aggressiveness, dominance, leadership, and political skill with, on the other, submissiveness, loyalty, and conscientiousness. One can have no doubt that there are genetic factors concerned as the basis of those differences. The same holds for the difference between high intellectual achievement and dullness or stupidity. It is in regard to the distribution of such qualities that I can see more important shorter-term genetic problems arising than at the level of medically significant genetic deterioration. Human mating groups have probably always existed—much of cultural anthropology has been concerned with their definition—and any genetic changes associated with current trends in civilisation will result from their influence on the composition of mating groups and the relative fertility of those groups rather than from any appearance and accumulation of deleterious mutations.

America has consistently led in initiating changes in human behaviour that eventually involve most of the world, and I have been interested to read hints from America of concentration of qualities in new types of mating group. The evidence suggests strongly that the group increasing in size most rapidly in the United States is the poorer stratum of Negroes with associated high fertility, low intelligence, and high crime rate. In an entirely different direction I have seen suggestions of a trend toward the development of mating groups of the intellectual élite in areas where there are high concentrations of intellectually demanding industrial or academic work. Modern technology, including the technology of administrative and managerial skills, requires a conscious effort to make the fullest use of genetically based intellectual quality. This has meant the provision of educational opportunity for young people almost to the limit of their capacity to absorb it, and the active search for talent. A college education has become in some sense a mark of an élite and is beginning to evoke signs of resentment in the other 85 per cent of the American population. One must expect that processes of this sort will lead to a variety of groups with special genetic capacity to excel in socially significant activities. Conceivably these may also become relatively closed mating groups distinctly analogous to the caste system of India. This might have curious, perhaps disastrous, social results in an affluent and largely leisured civilisation of the future.

Strictly speaking, there is only one definition of biological fitness. A group of organisms which has more offspring surviving to reproductive age is fitter than one which has less. It makes no difference what the factors are which determine that greater fertility. At the present time, by far the most important factor in determining the relative fertility of different groups is the degree to which various contraceptive techniques are used and their effectiveness. The differential application of birth control will have, or could have, greater influence on the biology of man than any other aspect of civilisation. We cannot forget that any circumstance which makes contraceptive methods easily available to the intelligent, mentally flexible, and socially responsible segment of the community, without also providing strong incentives for their use by the ignorant, superstitious, and irresponsible, will in a few generations change the whole character of our civilisation.

There may be something to be said by the sentimental for a Utopia in which simple peasants, full of all the homely virtues and of some comforting religious faith, raise families of 15 or 20. But every realist must see the mischievous futility of such sentimentality. I believe that we would be lacking in a sense of our responsibilities as scientists if we did not recognise the danger to everything which we cherish as intelligent human beings if opportunities for rational birth control are not made equally effective through all classes of all human communities.

I have confined myself almost wholly to questions of the changing distribution of human genotypes as influenced by factors associated with civilisation. Most of the papers will be concerned with more immediate impacts of the urban environment on the physical or mental health of human beings. This is a topic that I was once more enthusiastic about than I am now. My disillusionment springs from the failure of any significant action to be taken against the medical hazards of cigarette smoking. If we look at the vital statistics of western countries since about 1900 it is evident that civilisation, whether through preventive medicine, better medical care, or simple improvement in the standard of living, has had a striking effect in raising the average expectation of life. To the best of my knowledge the only environmental factor lethal enough to have a measurable effect in the opposite direction on vital statistics has been the cigarette. Since around 1950 it has become clear that cigarette smoking is almost wholly responsible for lung cancer, causes even more deaths from ischaemic cardiac disease, and is a highly significant factor in deaths associated with chronic lung incapacity. No other impact of civilisation has yet been shown to have so clear and measurable an effect in destroying man-years of potentially useful life. Poverty, worry, urban stress, marital disharmony, over-crowding, noise, air pollution, may all be responsible for medically or socially evil effects but none of them has been incriminated with the quantitative precision of the cigarette. I have been crusading against cigarette smoking in Australia since 1955. So have a large number of others, particularly Dr E. V. Keogh of the Victorian Cancer Council. I can see no response whatever from any government in Australia or from the trade union movement, and there has been no slackening in cigarette sales. There may be many other aspects of civilisation that have subtle but important harmful effects on human health and well-being. It is as well that we should know about them, but our complete failure in Australia and America to handle a measurably lethal situation of the first importance—the cigarette—must make us completely cynical about obtaining effective action against such things as stress, noise, air pollution, and so on. Our only hope seems to be in watching for new hazards to health and clamping down on them before vested interests arise to make action impossible. I suspect that marihuana *may* be much less objectionable than cigarette tobacco but we should get rid of it while that is still possible. There is perhaps just a hope that the outcry against sonic booms from the Concorde may deter the Americans from going on with their S.S.T. and spare the next decade from an intolerable nuisance. Once United States and Russian S.S.Ts. are in the air the situation will be hopeless. Most new industrial hazards are now relatively well controlled. One of the most remarkable achievements of preventive medicine has been the control of dangers in the use of nuclear reactors, atomic power plants, and weapons establishments. Here the danger was clear and readily measurable. The remedy was available from the start and has been built into the whole structure and function of the industry.

I should like this meeting to be something more than an academic exercise. I may well have overstated some of the points I have made and I am well aware that the accumulation of valid data with the full apparatus and integrity of academic scholarship is a necessary first step to action. But I should like to be able to hope even faintly that what we do in these two days will have some real impact on our future well-being in this country.

1 **Frances Barnes**

The Impact of Civilisation on the Biology of Man

The Biology of Pre-Neolithic Man

This symposium is concerned with the effects of civilisation on the biology of the human species and it is pertinent to introduce some comments on the biology of man before he was exposed to the conditions of civilisation. For this purpose we will somewhat arbitrarily date the beginning of civilisation from the time that a hunting-gathering way of life was superseded by one based firmly on agriculture, the domestication of animals, and a settled mode of existence.*

This change, which we shall refer to as the neolithic revolution, occurred first in the Middle East some 10,000 years ago and since then has taken place in almost all the inhabited areas of the world, so that today there are only a few vanishing peoples living by hunting and gathering. The process is of immense biological significance for mankind and has led, in only 400 to 500 generations, from palaeolithic hunter-gatherer societies to modern urban civilisation.

Before 10,000 years ago our ancestors had been hunter-gatherers for at least two million years. During this time man spread over much of the earth's surface but his effect upon the habitats in which he lived was relatively slight. Since the neolithic revolution he has modified his environment at an ever-increasing rate. Today he lives in, and must adapt to, conditions markedly different from those in which he made the transition from man-ape to man and in which he then lived for tens of thousands of generations.

The processes of civilisation, which were ushered in by the neolithic revolution, are based on man's capacity for culture and learned behaviour. These in turn depend to a great extent on his genetically-transmitted ability to acquire and use a symbolic language. This attribute probably developed in the first half of the Pleistocene epoch, during which time the volume of the hominid brain more than doubled, mainly because of growth of the neocortex.

Thus we find that the few remaining hunter-gatherer societies have

* We recognise of course that this process of change from a hunting-gathering economy occurred at different times in different places, that it may have been independently invented in different areas (e.g. in the New World) and that some societies have economies showing some, but not all of the traits mentioned in the text. Thus the Ainu had essentially a hunting-gathering economy combined with settled habitation, while the Siriono are essentially nomadic hunter-gatherers who plant a few crops.

well-developed cultures. P. V. Tobias,[1] writing of the Kalahari Desert Bushmen,* says:

'It is not the slow structural genetic adjustment nor the smooth and reversible functional accommodation but the swift, intelligent cultural adaptation which has permitted the Bushman to cope with the rigours of his environment. It is remarkable that even in a pre-neolithic economy like that of the Bushman, culture predominates over biological considerations in ensuring survival.'

Similarly, while Eskimos have certain genetic traits such as facial hairlessness and narrow noses which are advantageous in arctic conditions, it is nevertheless their complex cultural inventions that enable them to live as hunters on the border of the habitable world.

As these examples demonstrate, surviving hunter-gatherers live in the harsher habitats of the globe. We must therefore beware of attributing their characteristics unchanged to early man. However, data from these peoples, together with archaeological evidence and insights gained from man's primate relatives, are all that we possess, and we must do the best we can with these materials in our attempts to delineate the biological characteristics of hunter-gatherer man. Throughout this paper the interplay of traits possibly inherited from man's sub-human primate past and the increased flexibility of behaviour provided by his human brain will be stressed. In the analysis of the causes of behaviour we are still largely ignorant of the relative importance of the genetic as compared with the learned and culturally-transmitted components, but it is becoming increasingly apparent that the behaviour of the higher mammals, particularly that of the primates, is composed of complex trains of events in which the genetic and learned components interact in a sequential manner.

Among the primates only pre-agricultural man hunts to satisfy a substantial part of his dietary needs. It seems probable that, during the Miocene and Pliocene epochs in southern Africa, man's already omnivorous ancestors responded to the spreading savannah woodlands by increased hunting. This hunting was grafted on to a primate base, in which sight had for long been the dominant sense. Man became a mammalian oddity, a carnivore hunting by sight rather than smell.

Other primate traits played a part in his development. A tendency towards truncal erectness and climbing-by-grasping were pre-adaptive for bipedallism and the ability to hold, manipulate, and carry objects in the freed hands. Primate stereoscopic colour vision and high exploratory drive contributed to the matrix from which he emerged. Later the acquisition of language provided a basis for the development of abstract thought and the expansion of culture so that man was able to amplify the social structure he inherited from his primate ancestors

* Material on the Bushmen used in this paper has been taken mainly from the following works:

G. B. Silberbauer, *Bushman Survey*, Gaberones, 1965, Bechuanaland Govt; R. B. Lee, *Subsistence Economy of the !Kung Bushmen*, University Microfilms, 1967, Ann Arbor; E. M. Thomas, *The Harmless People*, London, 1959, Martin Secker and Warburg.

with organisation based on concepts such as kinship, exogamy, and incest. Language also facilitated the construction of ideologies and the development of a sense of time and personal identity—traits which are, to a marked degree, characteristic of the human species.

Hunting stimulated the use of tools and weapons, which led in turn to better hunting and eventually to the making of tools 'to a set and regular pattern'. This achievement has been regarded as a hallmark of man, even if rudimentary tool-fashioning can be seen in chimpanzees when they break twigs to a suitable length in order to get ants out of holes.

The role of hunting behaviour as a determining factor in man's evolutionary development raises interesting speculations. W. S. Laughlin[2] sees it as 'the master behaviour pattern of the human species—the organising activity which integrated the morphological, physiological, genetic and intellectual aspects of the individual human organisms and of the populations which comprise our single species'. Early vegetarian ape-men such as Paranthropus died out: an omnivorous ape-man survived.

It has been suggested in particular that hunting favoured the development of economic co-operation amongst males and food sharing within the group. Other killers of large prey such as wolves show the same characteristics. A big kill, made possible by pack behaviour, is used to the best advantage of the group if it is shared. Interestingly, Jane Goodall[3] reports an instance of real economic co-operation among chimpanzees when two combined to hunt and kill a colobus monkey. She also describes 'begging' gestures for meat which are sometimes rewarded. This contrasts with primate frugivorous feeding which is a solitary affair. The existence of a fruiting tree may be signalled vocally but, on arrival, each animal selects its own food, prepares it if necessary and eats it.

Hunting may also have selected for the development of efficient eye/hand co-ordination, good vision, and acute hearing. R. H. Post[4] has demonstrated a falling-off in the level of visual acuity and an increase in red/green colour blindness paralleling the removal, in time, of a population from the hunting-gathering way of life. Bushmen have heard a single engined Cessna aircraft while it was still seventy miles from the airstrip and many of them can see the four moons of Jupiter with the naked eye. However, before attributing the changes in urban society entirely to relaxation of selection pressures we would have to consider whether modern living, with its increased level of noise and very different use of the eyes, is having physiological effects within the lifetime of individuals.

At some stage the sexual division of labour became general. In all extant hunter-gatherer societies men are the hunters of large animals even if, as with the Pygmies, women play an integral part as beaters. That this was so by the end of the palaeolithic era in Europe is shown by the hunting scenes of cave art. Although there is no evidence that early hominid females did not hunt, as the human species evolved it was the males who became the hunters; females were increasingly burdened by the care of the young. The enlarging human brain was associated with increasing immaturity at birth and, in order to exploit the flexibility allowed by this complex brain, infants developed increasingly slowly. As fur disappeared, perhaps because of the advantages of quick heat loss when hunting in a hot climate,[5] the young could no longer cling but had to be carried or tended in a home base. Women would have been inefficient hunters of large game.

Later, as humans colonised colder areas, they had to compensate for the loss of fur by such cultural devices as clothing, fire, and efficient shelter. For the last we can again see a possible phylogenetic basis; among other primates, apes are the only nest builders.

Looking at modern hunter-gatherer groups we find that diets are extremely varied. Eskimos are almost entirely carnivorous as there is little vegetable food available, while for the Hadza of Tanzania meat provides only 20 per cent by weight of the food eaten and the women's gathering supplies the subsistence base.[6] Both groups have been described as healthy. Amongst four Australian Aboriginal groups living in the bush in 1948, meat and/or fish provided from 37 to 97 per cent of the diet. However, two men from the band living almost entirely on meat became tired of the diet and walked to a mission for supplies of rice and flour.[7] Chimpanzees when eating meat take a handful of leaves between mouthfuls.

Recent studies have shown that hunter-gatherer economies can provide adequate and relatively secure food bases. The time spent hunting or gathering by Bushmen and Aborigines, though highly variable, averages three to five hours a day.[8] In Bushman society most of the day is spent resting, talking, visiting, and making weapons or other artifacts; old people are fed and cared for, and children do not have to assume adult economic responsibilities at an early age. The Bushmen say 'why should we plant when there are so many mongongo nuts in the world?' If there is no food in a Hadza camp at nightfall everyone knows more can be gathered tomorrow. However, in some habitats the food quest is more frustrating. Among the Siriono, in the Bolivian rain forest, there appears to be a good deal of hunger. This leads to widespread anxiety about food, and often to aggressive behaviour.[9]

In hunter-gatherer societies everyone is involved in the physical effort required to obtain food. This, combined with the fact that the food has a higher fibre content than that of a modern diet, contributes to the finding that in hunter-gatherer societies, as amongst wild animals, obesity is virtually unknown. However, a liking for sweet things is an ancient trait which man shares with other primates, and which is adaptive in a frugivorous life. A Spanish cave painting, late palaeolithic or early neolithic in date, shows men using ropes to reach honey, much as the Veddas did in the early years of this century. Honey is also prized by the Pygmies, Bushmen, and Aborigines, and if more of this highly concentrated carbohydrate had been available in the environment, obesity and widespread dental caries might have become problems at an earlier date.

Emotional overtones are attached to meat, even in groups where it is not the mainstay of life. For the Hadza and Xavante it is 'real' food, and the symbolic food exchanges of the latter group are made with meat pies.[10] Bushmen evaluate plant foods rationally and they impose few taboos on their use; in contrast there are taboos of varying strictness on all animal foods. Is this because emotional overtones became attached to meat through hunting and killing or is it because of a greater tendency in men as opposed to women to ritualise their activities? At all events, by the end of the palaeolithic era in Europe we see cultures, reflected in the cave art, where men hunted large animals and in which totemism, hunting magic, and fertility rites probably existed much as they exist in some cultures today.

Let us turn to the demographic picture presented by hunter-gatherer populations. Modern palaeolithic man lives in relatively small face-to-face groups and we can safely assume he has always done so. Perhaps from early times there were temporary larger gatherings made possible by seasonal abundance of food and used for the interest and excitement of extended social intercourse. Such gatherings are very much a part of Australian Aboriginal and Eskimo life. In extreme conditions we find the subsistence unit reduced to the nuclear family, but, when things improve, the foraging band, a fundamental group of palaeolithic society, reasserts itself. These bands appear to vary around a modal size of twenty-five to fifty people. Both ecological and social factors may influence the size of bands but our knowledge of their operation is incomplete and imprecise. The nature of the food base and the techniques which have been evolved to obtain it both contribute. Among the Bushmen, G/wi bands have around fifty members while !Kung bands usually contain ten to thirty people, with much visiting, joining up and redividing taking place. Among the Pygmies, six to seven nuclear families are needed to form an efficient hunting unit and the Australian Aboriginal foraging band consisted of twenty to fifty adults. B. J. Williams[11] suggests that thirty-five to seventy-five people are needed to maintain exogamous relations with surrounding bands and even if the Vedda hunting unit had only one to two extended families their land-holding groups were often larger. The usually endogamous band of the Siriono was made up of three to five extended families totalling about seventy people. It seems that both cultural and ecological factors establish upper and lower limits of band size.

An important aspect of most hunter-gatherer bands is that they are less closed than the groups of most other primates. Some band-changing by gorillas and baboons (almost always males) has been observed, while the details of the much looser chimpanzee social organisation are still not clear. Itani and Suzuki[12] have analysed three large chimpanzee groups (thirty to fifty animals) and found fewer young adult males than would be expected on the basis of observed sex ratios among infants and juveniles. They suggest that this age group keeps 'some social distance from the large sized groups', perhaps in part because of conflict between mothers and their near adult sons as these latter achieve positions of dominance over females. The tendency of young males to roam further afield may also contribute. These authors even speculate as to whether male chimpanzees normally change groups as they become fully adult but state that further work is required on the relations between the fluid small bands of chimpanzee society and the larger groupings which occur from time to time.

In the human species individuals may be *required* to change groups —on marriage. Thus hunter-gatherer bands, unlike those of other primates, are units of a wider society, which is held together by kinship bonds.

We should note here that although the links in this network of bands could have stretched far, for any individual human being almost all contacts with members of his own species would be with persons well known to him. The stresses, but also the stimuli, involved in new contacts would have been few. It has been estimated that 400 to 500 people could have been the total number of persons met in an average lifetime. On the other hand we must also allow for the general fear of the unknown stranger, which may have been much more acute when he *was* encountered, than it is in our more sophisticated society.

From his work in Australia, and supported by data from the Shoshoni and Andamanese, Joseph Birdsell[13] has postulated that there is a tendency among generalised hunter-gatherer populations for the linguistic unit (or tribe) to approximate 500 people, even if in some cases there may be a considerable variation. He suggests that this figure may be largely the result of forces operating within the communication system provided by symbolic speech and foot mobility, but also notes that a change in size may follow a change in social mores.

Population densities varied with the ecology of the habitat, but must in all cases have been low in comparison with those occurring after the neolithic revolution and almost infinitesimal when compared to the densities found after the industrial revolution. As modern work with mammals continues to reveal profound physiological and psychological changes stemming from overcrowding, such as increased rates of aggression, increased parasite load and decreased resistance to infection, impaired reproductive performance and abnormal maternal behaviour, the question whether such factors are operating in our modern urban environments becomes a valid one. Later we shall note the increasing tendency towards stratified hierarchical human societies after the neolithic revolution. The struggle to survive and reproduce in more crowded conditions may have promoted aggressive and dominance behaviour patterns in a way which was unnecessary (or even harmful in terms of the ability of the band to survive) in the hunter-gatherer way of life. In this connection it is interesting that there is an increased value in being dominant in crowded rabbit populations.[14]

Henri Vallois concludes from fossil evidence that in prehistoric times few people lived beyond forty and most females were dead by thirty. In a Neanderthal series 50 per cent of the people had died before twenty.[15] The picture, based admittedly on scrappy evidence, is of early death, especially of females, and moderate childhood mortality. Parents would not have been able to care for the later-born children throughout their childhood. In contrast, at the present day the expectation of life at birth has risen to seventy in some Western societies.

Even in contemporary hunter-gatherer societies there are difficulties in filling in the demographic picture. C. B. Silberbauer thinks the G/wi in the Kalahari Desert age rapidly and few live beyond forty-five. On the other hand George Grey,[16] on the basis of genealogies and observations obtained on his journeys in 1837-9, is certain that Australian Aborigines often lived to be seventy. Amongst the Xavante, a South American Indian group still predominantly nomadic when studied in 1962, most men were dead by forty-five, but homicide is thought to have played a part here as the older men struggled for power.

It has been estimated that births to chimpanzee and gorilla mothers are often spaced at three- to four-year intervals. This also appears to be the case in many hunter-gatherer groups. The average number of children born alive to a G/wi woman is three, spaced at four- to six-yearly intervals in the years from eighteen to the early thirties. Grey found 4·6 children per completed Aboriginal family associated with a high infantile mortality rate and probably frequent miscarriages, perhaps sometimes induced. Semang women are said to bear a child every three years. We are largely ignorant of the mechanisms by which these spacings are achieved. Physiologically the suppression of ovulation by lactation may be important. In the human species cultural factors supplement any physiological mechanisms which may be operating. Thus abstinence, abortifacients, oral contraceptives, and infanticide are all reported in the literature, but except for the last we have little real evidence as to the extent of their use or of their efficiency. It appears that an appreciation of the fact that too frequent child-bearing jeopardises the chances of survival of mother and child, acts, under hunter-gatherer conditions, as a stimulus to population control. Mary Douglas has suggested that after the neolithic revolution population homeostasis still occurs but that 'the kind of relation to resources that is sought is more often a relation to limited social advantages than to resources crucial to survival'.[17] Rather than an appreciation of the resources of the habitat, the social structures of the group and a belief that cultural prizes are within one's grasp become the important mechanisms for regulating population. If correct, this again demonstrates the effect of civilisation in decreasing the pressures of the natural habitat and increasing the importance of cultural factors.

Some animal species limit population growth by means of territorial organisation and only those individuals in possession of a territory breed. However, man does not relate himself to land in a rigidly territorial fashion and a breeding pair of humans does not have to possess a den or a territory before mating can occur. Like gorillas and perhaps chimpanzees, hunter-gatherer groups often operate flexible, overlapping home range systems. Each group normally hunts and gathers over an area of land to which it has a special relationship but the boundaries of which are frequently ill-defined and not rigorously defended. In different habitats and under different ecological and social pressures access to land other than that over which the group normally forages is more or less easy. Such access is often regulated by social mechanisms based on kinship. Patterns vary from that of the Siriono where a band simply moved away if it came across evidence that another band was hunting in the area, to the complicated land use patterns of the Australian Aborigines. These have been analysed by W. E. H. Stanner,[18] who concluded that the ecological life space had two components, an estate and a range. The estate was the 'ground' or 'dreaming place' and was important as the ritual focus of a patrilineal descent group. This descent group formed the nucleus of a foraging band which hunted and gathered over a range. Estate was usually but not always contained within the range. Groups often went outside their ranges for social intercourse, particular foods or raw materials and, especially in the harsher environments, ranges often overlapped.

For the Dobe Area Bushmen in 1964, Richard Lee mapped nine territorial areas and pointed out that each territory had enough food to sustain some members of the band which normally foraged over it throughout the year but that arrangements were flexible enough to 'permit people to collect and eat freely in territories other than their own'.[19]

Eskimos also had, in general, little idea of owning hunting territories although Alaskan Eskimos, perhaps influenced by their Indian neighbours, had more developed ideas of property rights in land.[20] The Veddas also seem to have had well-defined and defended territorial borders.[21]

In general, however, I think we must conclude that it was the neolithic revolution, with its attendant necessity to protect stock and crops and the greater opportunities it provided for accumulation and storage of goods, which set the stage for many of the acquisitive and territorial wars which disfigure the subsequent history of mankind.

Let us look more closely at the social organisation of foraging bands. Many non-human primate groups appear to be held together by bonds based on the three biological categories of male, female, and young, giving six possible two-way bonds. As more long-term studies of primate groups become available we are finding that even in non-human groups kinship may be acting as a primary bonding mechanism.[22] In human groups we have already commented on the importance of the *concept* of kinship as an organising principle in society—thus extended kinship classification often covers all members of the band and members of other bands as well.

All-male groups are a prominent feature of contemporary human societies. We can speculate whether the bonding mechanism between males in early human societies was based on an aggressive dominance hierarchy,[23] as in baboons, which later was modified by the need to co-operate in hunting, or whether the pattern was one of mutual attraction, as appears to be the case with chimpanzees.[24] In favour of the latter it would seem that in modern hunter-gatherer societies there is in general a lack of hierarchical structuring.

Thus, for the Bushmen, Silberbauer makes it clear that no adult member of the band has more authority than any other by virtue of superior status or strength, and of the Pygmies Turnbull writes 'There were no chiefs, no formal councils. In each aspect of Pygmy life there might be one or two men or women who were more prominent than others, but usually for good, practical reasons'.[25] The co-operative mores of the Diomede Island Eskimos were noted in 1899, as was their lack of an executive head.[26] In the New World the area where a complex social system of status and rank developed was Meso-America, which also possessed intensive irrigation and farming.[27]

Male-female bonds are stressed in human societies where much social organisation centres round relatively permanently mated pairs. This is in contrast to many primate societies where heterosexual bonds tend to be ephemeral, operating only when the female is in oestrus. It is interesting that relatively permanent mated pairs probably occur in another co-operative hunter—the wolf. Long-term apportionment of females may help damp down aggressive impulses between males in both species. The disappearance of oestrus in the human female and her capacity to participate in sexual intercourse at all times in the reproductive cycle could have reinforced the heterosexual bond and provided a basis for the emerging nuclear family which became the major child rearing unit of the human species.

Female-female bonds are less in evidence but groups of females and young have been noted amongst chimpanzees, and women go gathering in groups, perhaps for company, perhaps partly for security. Pygmy women working in the village do so in twos and threes, again for company. Whether this represents a bond between females or whether it merely reflects a dislike of being alone is perhaps open to question. In either case the isolation felt by the suburban housewife may rest on a firm biological basis.

The young of primates often appear to attract the males. Hall and DeVore have noted such behaviour in baboons and Schaller comments on the tolerance of adult male gorillas towards juveniles. Bonds between females and young are well-developed for obvious biological reasons. The attractiveness of infant rhesus monkeys to females other than their mothers has been described,[28] and for chimpanzees Jane Goodall has recorded a situation where an older sister 'adopted' an infant whose mother disappeared.[29] In this case kinship, as manifested in a sibling relationship, may also have played a part, for the older brother on occasion protected the baby. A strong mother-infant tie is necessary for successful rearing, as the young primate is not left in a den but must go everywhere with his mother. Much has been learned about the cues which release maternal behaviour in primates including humans, and it is becoming increasingly obvious that such behaviour is not entirely instinctual. Adequate adult maternal and sexual behaviour both depend in part on adequate socialisation experiences during development. It is interesting that socially deprived chimpanzees can learn the adult sexual pattern more effectively than deprived monkeys.[30] Deprived human children can probably learn even better than chimpanzees but some Western societies may have relied too much on this capacity, thereby contributing to the level of sexual maladjustment in their adult populations.

J. P. Scott has put forward the idea that there exist critical periods in development, one of which he calls primary socialisation.[31] During this period bonds are formed and patterns of behaviour develop which contribute to adult social organisation. Thus in dogs and wolves primary socialisation occurs after locomotion is established and while adult feeding patterns are developing. The mother is away from the den much of the time and primary socialisation occurs between litter mates, forming a good basis for later pack behaviour between age-mates. In humans primary socialisation occurs during the period stretching from six weeks to six months, before walking or adult feeding patterns are established and while the infant is still totally dependent. The bonds created are, therefore, between the infant and the caretaker; this is usually the mother, perhaps supplemented by the father and siblings. Scott notes that there is a basis here for the development of inter-generational bonds in human societies and for the experience of welfare workers that the best time for a baby to be adopted is during the few weeks after birth before the period of primary socialisation begins.

Bonds between young are important in primates and are expressed in social play.[32] In rhesus monkeys adequate peer group experience is more important than adequate mothering for the development of effective adult sexual behaviour.[33] The Israeli kibbutzim use the peer group as the main socialising agent. Mother-infant contacts in the first six months are intermittent, mainly for feeding, and most of the baby's care is relegated to nursemaids. After infancy the child lives for most of the time in a peer group society under the control of a succession of caretakers and teachers. The finding that 'the sabra [one who has grown up in a kibbutz] as a personality type stands impervious to integration in social forms other than those beginning and ending with his peer group'[34] gains new significance in the light of Scott's suggestion that it is the bonds formed between mother and infant during primary socialisation which may provide a basis for inter-generational bonds in human societies. It is also interesting that many kibbutzim are now allowing mothers to take a greater part in the care of their young babies.[35]

Supplementing such bonding mechanisms as these, human groups are held together and differentiated by a variety of specifically cultural concepts based on the appreciation of kinship relations. Incest taboos, and rules of exogamy and/or endogamy are to be found in most human groups. The origins of these practices are obscure, and different hypotheses, not all mutually exclusive, have been put forward to explain them. Basic physiological and psychological reactions may play a part. Biological considerations such as the dysgenic effect of very close mating may have favoured groups prohibiting it, and the social advantages of kin links with surrounding groups may have helped the development of rules of exogamy.[36]

Let us turn to a consideration of the health of hunter-gatherer bands. We have a good deal of evidence to suggest that at least in some habitats their members achieved a high level of physical fitness, and lived in effective symbiotic relationship with their surroundings. Feats of endurance are often mentioned in the literature. Thus an Aboriginal messenger could cover fifty miles in a day and a Bushman walk thirty-four miles in five hours in heavy sand on a hot summer day. !Kung women carry their own weight in firewood.

Blood pressure in both Bushmen and Aborigines is low and does not increase with age though there are indications that it may increase with acculturation.[37] In spite of this, widespread arteriosclerosis has been found in Bushmen, which may help to account for their tendency to premature ageing.[38] Nye found no arteriosclerosis in a group of old Aborigines, many of whom he thought were seventy years old.[39]

The nomadic way of life protected hunter-gatherers to some extent from diseases characteristically occurring in conditions of poor sanitation. However, the habitat may have been equally important, dry tropical conditions being more inimical to most micro-organisms than wet tropical ones. Dunn[40] has demonstrated the increasing number of pathogenic helminths and protozoa occurring in hunter-gatherer populations as one moves from dry arid habitats to tropical rain forests. The principal diseases suffered by the pre-contact Siriono of Bolivia in their tropical rain forest are thought to have been malaria, dysentery, hookworm, and skin diseases. In this group disability and mortality seem to have been considerable and fear of illness widespread, as seriously ill adults were sometimes abandoned if the group had to trek to another food area.

The relative isolation of groups and their small size meant that infectious diseases producing long-lasting immunity could never become endemic within the group. The disastrous effect of the introduction of such diseases as measles, chicken-pox, and whooping cough into isolated populations after contact with Western civilisation is circumstantial evidence that these populations had not previously had experience with these pathogens during the lifetime of their members.

The position with regard to degenerative and inflammatory diseases is rather different. From the bones the archaeologists unearth we find ample evidence for widespread degenerative bone pathology; Neanderthal man fairly frequently suffered from osteoarthritis of the jaw and spine and there is a series of seventeen adult arthritic skeletons (out of twenty-seven) at a Middle Horizon site in California.[41] This group of people got most of their food on the shores of Bodega Harbor, and the skeletons demonstrate the close association between disease and the way of life followed by a population. The series also produced two males with ostcomyelitis and two with sinusitis–while sinusitis is also found in palaeolithic skulls from a cave at Lozère. Australian Aborigines seem to have suffered from a peculiar type of periostitis and osteitis resulting in the condition known as 'boomerang legs'. Although dental caries is rare in hunter-gatherer groups alveolar abscesses were common in the Californian skeletons. We must, I think, conclude that from early times inflammatory and degenerative bone pathology may have been important causes of disability.

Traumatic bone injuries are found both in archaeological material and amongst living hunter-gatherers; healed fractures are common. In many cases the general hazards of a hunting life are responsible. Falls from trees produce injuries amongst the Siriono, fractured arms and legs are found in Aborigines hunting over boulder strewn country while Colles's fracture and fractures of the small bones of the leg occurred in the Californian men gathering seafood on a slippery shore. A *Homo erectus* femur shows an exostosis, probably the result of trauma. In some cases injuries may have been caused by intraspecific violence. Thus Australian Aboriginal skeletons have been reported as having a high incidence of 'parry' fractures of the forearm.[42]

Some environments provide other hazards. Drowning is a frequent cause of death among young male Eskimos while eye injuries from branches are common among Bushmen. Snake bite and attack by predators are significant dangers in some habitats; the latter may have been even more important in early savannah environments. Wounds in Aborigines and Eskimos have been reported as healing particularly well. How much this is an expression of their fitness and how much it is due to the relative sterility of their extreme environments is open to debate.

Deficiency diseases are rare in palaeolithic groups, as one would expect on biological grounds. Except in difficult seasons, hunter-gatherer diets are adequate for good health even in relatively unfavourable habitats. Worm infestation is no doubt as old as man and he has lived on meat long enough to become the definitive host for two types of tape worm. However, under natural conditions a symbiotic relationship may have existed, causing little disability to the host.

There is some evidence that, in hunter-gatherer groups, carcinoma is rare and the same is true for coronary thrombosis. However, the relatively young age at death introduces problems in evaluating the significance of this evidence. The incidence of congenital abnormalities probably varied from group to group, perhaps influenced by the amount of inbreeding. In one inbred Siriono group club-foot has reached 15 per cent.

Eye diseases, for example trachoma (probably pre-contact among the Aborigines) and conjunctivitis (widespread among Bushmen), could have existed in early groups, and mycotic skin conditions are in the same category.

Thus our final picture, hazy as it is, suggests that the health of ancient palaeolithic man varied, as we would expect, in the different habitats in which he settled. In general, the hunter-gatherer way of life provided a diet which kept him in good health and the demands of his nomadic life kept him in good physical condition. However, in some environments, and with some modes of food getting, wear on joints led to widespread osteoarthritis, as well as to other inflammatory and traumatic bone pathology. Infective diseases producing immunity were probably rare, but other infective illnesses such as gastro-enteritis, possibly caused by the same organisms as produced the osteomyelitis, could have contributed to a high infantile mortality. In this context we may note that G. B. Schaller considers disease to be probably the chief cause of death in the gorilla, respiratory disorders, helminths, and micro-filaria being indicted. One adult gorilla definitely died of gastro-enteritis during the study and the male mortality rate prior to puberty was calculated to be 47 per cent. We can only guess at the reasons for the relatively short expectations of life of most of the population in many hunter-gatherer groups. In pleistocene populations problems of pregnancy and parturition in a relatively recent bipedal species which was producing an infant with an unusually large head perhaps contributed to the earlier age at death of women as compared to men. This statistic has been reversed in modern Western populations where women in general outlive men. We are even more ignorant about the mental health of hunter-gatherer groups. Silberbauer did not observe any 'neurotic symptoms' amongst the G/wi Bushmen. Ehrstom,[43] investigating Greenland Eskimos in 1948/9, suggested that the incidence of psychosomatic disease increased with acculturation while hysterical manifestations, such as 'Kayak dread', were more often found in those with less experience of Danish culture. Abbie[44] thinks that pre-contact Australian Aborigines were an adaptable and 'on the whole a happy' people, while Turnbull depicts the Congo Pygmies as resilient, happy, and well-integrated into their forest environment. However, it is extremely difficult to gather firm evidence concerning the psychological condition of hunter-gatherer populations.

One major subject we must consider is the question of intraspecific aggression in the human species. Man holds the unenviable position of being the species most prone to behaviour which actually results in the killing or serious wounding of a member of the same species. We can assume that, like his primate relatives, man's ancestors had, at some stage in history, a way of life in which the use of aggressive behaviour was integrated into the total life pattern of the species in such a manner as to prevent the killing of conspecifics. Thus the problem is to define the forces which brought about a change in behaviour and resulted in the fairly frequent killing of members of the same species.

Predation obviously involves aggressive behaviour patterns and favours the development of organs capable of inflicting damage, such as canine teeth and sharp claws. Animals thus armed must, if the species is to survive, ensure that they do not consummate aggressive behaviour patterns against conspecifics. This has been achieved by such means as ritualisation of fighting behaviour, submissive gestures which inhibit the stronger animal from pressing home the attack (dogs, wolves), or the primary defence of territory by aggressive behaviour so that once the weaker animal has fled the behaviour subsides (deer). Therefore we find in many species aggressive behaviour patterns which inflict minimal intraspecific damage but which help in the food quest, play a part in social organisation (territorial and dominance behaviour) and are essential components of sexual behaviour. This latter aspect of human aggressive behaviour has received much attention from psychiatrists recently. MacLean has demonstrated in the spider monkey that penile erection is part of both the aggressive-dominance behaviour pattern and the courtship pattern:

> 'From the level of the anterior commissure caudally one can follow neural structures involved in anger and fear and combative behaviour lying next to those concerned with feeding and sexual response.'[45]

In common with other predatory animals man developed his armamentarium, but it was an armamentarium with a difference. It consisted of weapons and was based on primate motor patterns and not on the development of claws or canines. Agonistic throwing and clubbing have been reported in the great apes and in the case of throwing both frequency and accuracy of aim appear to be higher in males than females.[46] This may be yet another primate trait which contributed to the sexual division of labour in the human species.

The use of weapons enabled palaeolithic man to make a success of his change to a more carnivorous diet. However, such predation, with its associated aggressive behaviour, has been, in evolutionary terms, a late development and, at least in its later stages, a rapid one. Perhaps, like the present day gorillas and chimpanzees, man's sub-human primate ancestors did not depend, to any great extent, on the use of aggressive behaviour to maintain their social organisation. In this case the necessity for highly effective instinctive checks on intraspecific aggression would not have arisen.

It is difficult, though possible, for one unarmed man to kill another. With the development of weapons the position changed, and checks which had been adequate may no longer have been so, especially with weapons capable of use at a distance. However, we should not overlook the possibility that quite a strong selection pressure in favour of the development of effective inhibitory mechanisms may have been operating throughout the period in which weapons were developing.

It is not easy to decide how far back in time we can find reliable proof of intraspecific killing. Evidence of death following a fracture of the skull has been found as early as the Australopithecines but there is nothing to indicate whether the fracture was the result of a deliberate blow or due to an accident. If it was deliberately caused it may have been inflicted either by a member of the same species or by a member of one of the other contemporary hominid species. *Homo erectus* and Neanderthal skulls have been found with damaged bases, suggestive of cannibalism. Some also show fractures in the temporal region, indicating that their owners probably died violent deaths. Opinions differ as to whether these finds constitute certain evidence of intraspecific killing or whether they only point to ritual necrophagy as part of funerary rites.[47] If they are evidence for the deliberate killing of fellow men for ritual purposes it suggests that a concomitant of becoming human and developing the power of symbolic thought was a loss in effectiveness of whatever innate inhibitory mechanisms were in existence. It seems unlikely that at this stage in the development of the human species, when in many environments predators and accidental death must still have been considerable hazards, the species could have survived intraspecific killing on a large scale. It is, however, possible that the benefits derived in the development of learned and culturally transmitted behaviour outweighed the survival disadvantage to the species of a certain amount of intraspecific ritual killing which was integrally linked with the process of humanisation. Thus we cannot rule out of court the possibility that intraspecific killing has an ancient history.

As food-providing techniques increased in efficiency, populations grew denser and groups larger. There is evidence that at about the same time there may well have been an increase in intraspecific violence.[48] Thus widespread violence as well as dominance-structured societies may both be in large part human reactions to the processes of the neolithic revolution. Crowding, as has already been mentioned, increases the rate of aggression in many mammal communities.

As far as contemporary hunter-gatherer societies are concerned the killing of fellow humans certainly occurs. Sometimes the victim is a member of the group, more often he belongs to another band. Most killing seems to be on a small scale and is often over women, but more organised warfare occurs in some areas, as for instance in N.E. Arnhem Land in Australia.

However, to the Bushmen aggression is unthinkable and they have been called the 'harmless people'. In this society young children are reported to be about as aggressive as young European children and the aggression goes unchecked unless injury is about to result. In middle and later childhood aggression is not permitted.[49] It is probably significant that the models in the adult world who will not tolerate aggressive behaviour are themselves non-aggressive. In the USA it has been found that a similar pattern of child rearing is the one most likely to produce non-aggressive children.[50]

This example illustrates the part played by learning and the cultural mores in the control of aggression in human societies—a development also made possible by the humanisation of man and the increased control of impulse by learned behaviour associated with the enlargement of the neocortex. This is not to deny the possibility that a genetic component may be operating, but the observation that young Bushmen are as aggressive as young Europeans would suggest that genetic factors are in this case of less importance than cultural factors.

We can conclude with a few final points. Man is a diurnal animal and until he tamed fire (Peking man half a million years ago) he probably slept throughout the hours of darkness much as the apes do today. Perhaps like them and like some of our contemporaries he also took a routine siesta. Changes in various physiological processes are evoked by changing the sleep pattern and the amount of exposure to light. The same is true in relation to noise and eating patterns. Urban man lives in a noisier world than did his palaeolithic ancestors and probably spreads his food intake more evenly throughout the day. Alterations in carbohydrate metabolism are known to occur when food is concentrated in one meal a day or spread throughout waking hours.[51] Whether the patterns of behaviour now expected from civilised man in such areas as these are within the limits of physiological adaptability of the species or whether and in what conditions they contribute adverse stresses require further study.

The few remaining groups of hunter-gatherers will soon have become agriculturists or industrialists. The way of life which has sustained the human race for 99 per cent of its time on earth will have disappeared. Man will live in environments largely created by himself, and often spectacularly different from those of his palaeolithic ancestors.

Unless these environments provide for the biological needs stemming from his primate inheritance and hunter-gatherer past, the resulting stresses may prove too much for him notwithstanding his unique ability to adapt to changing conditions through cultural processes.

Summary

In this paper recent studies of non-human primates and contemporary hunter-gatherer societies are considered, along with archaeological data, in an attempt to delineate some of the biological characteristics of pre-neolithic man. Behaviour based on these characteristics formed the springboard from which man took off when he embarked on the changes of the neolithic era, a brief 10,000 years ago. Developments stemming from the introduction of agriculture, the domestication of animals for food, and a relatively settled mode of existence have profoundly altered both biological and sociological aspects of man's way of life. Too few generations have passed for there to have been much genetic change and man is facing the space age with a genetic equipment determined largely by the selection pressures which produced his pre-neolithic ancestors. Clearly some knowledge of the biological and sociological attributes of the pre-neolithic way of life is a prerequisite for understanding the effects of civilisation on the biology of man. The paper therefore covers such topics as the role of hunting behaviour in human development, the importance of the capacity for culture, the health and demography of hunter-gatherer groups, the evolution of social behaviour, and the problem of human intraspecific aggression.

References

1 P. V. Tobias, 'Bushman Hunter-gatherers: A Study in Human Ecology', in *Ecological Studies in Southern Africa*, D. H. S. Davis (ed.), The Hague, 1964, Dr W. Junk, 67-86.

2 W. S. Laughlin, 'Hunting: an Integrating Biobehaviour System and its Evolutionary Importance', in *Man the Hunter*, I. DeVore and R. B. Lee (eds.), Chicago, 1968, Aldine, 304-20.

3 J. Goodall, 'Chimpanzees of the Gombe Stream Reserve', in *Primate Behavior*, I. DeVore (ed.), New York, 1965, Holt, Rinehart and Winston, 445.

4 R. H. Post, 'Population Differences in Vision Acuity: A Review, with Speculative Notes on Selection Relaxation', *Eugen. Q.*, **9**, 1962, No. 1, 189-212; R. H. Post, 'Population Differences in Red and Green Colour Vision Deficiency: A Review, and a Query on Selection Relaxation', *Eugen. Q.*, **9**, 1962, No. 1, 131-46.

5 W. La Barre, a comment (p. 148) on 'The Human Revolution' by C. F. Hockett and R. Ascher, *Curr.Anthrop.*, **5**, June 1964, No. 3, 135-68.

6 J. Woodburn, 'An Introduction to Hadza Ecology', *Man the Hunter*, I. DeVore and R. B. Lee (eds.), Chicago, 1968, Aldine, 49-55.

7 M. McArthur, 'Food Consumption and Dietary Levels of Groups of Aborigines Living on Naturally Occurring Foods', *Records of the American-Australian Scientific Expedition to Arnhem Land*, Melbourne, Vol. 2, 1960, Melbourne University Press, 127.

8 F. D. McCarthy and M. McArthur, 'The Food Quest and the Time Factor',

Records of the American-Australian Scientific Expedition to Arnhem Land, Melbourne, Vol. 2, 1960, Melbourne University Press, 147.

[9] A. R. Holmberg, *Nomads of the Long Bow,* Washington, 1950, Smithsonian Institution, Institute of Social Anthropology, Pub. No. 10.

[10] J. V. Neel *et al.,* 'Studies on the Xavante Indians of the Brazilian Mato Grosso', *Human Genetics,* **16**, 1964, No. 1, 52-140.

[11] B. J. Williams, 'The Birhor of India and some Comments on Band Organization', in *Man the Hunter,* I. DeVore and R. B. Lee (eds.), Chicago, 1968, Aldine, 126-31.

[12] J. Itani and A. Suzuki, 'The Social Unit of Chimpanzees', *Primates,* **8**, 1967, No. 4, 355-81.

[13] J. B. Birdsell, 'Some Predictions for the Pleistocene based on Equilibrium Systems among recent Hunter-gatherers', in *Man the Hunter,* I. DeVore and R. B. Lee (eds.), Chicago, 1968, Aldine, 229-40.

[14] K. Myers, 'The Effects of Density on Sociality and Health in Mammals', *Proc. ecol.Soc.Aust.,* **1**, 1966, 40-64.

[15] H. V. Vallois, 'The Social Life of Early Man: The Evidence from Skeletons', in *Social Life of Early Man,* Viking Fund Publications in Anthropology, Vol. 31, S. L. Washburn (ed.), New York, 1961, Wenner-Gren Foundation, 222.

[16] G. Grey, *Journals of Two Expeditions of Discovery in N.W. & W. Australia, 1837, 8, 9.* 2 vols., London, 1841, T. and W. Boone, Vol. 2, 247.

[17] M. Douglas, 'Population Control in Primitive Groups', *Brit.J.Sociol.,* **17**, 1966, No. 3, 263-73.

[18] W. E. H. Stanner, 'Aboriginal Territorial Organisation: Estate, Range, Domain and Regime', *Oceania,* **36**, 1965, No. 1, 1-26.

[19] R. B. Lee, *Subsistence Economy of the !Kung Bushmen,* University Microfilms, 1967, Ann Arbor, 138.

[20] E. M. Weyer, *The Eskimos,* Hamden, Connecticut, 1962, Archer Books, 124; K. B. Smith, *The Eskimos,* London, revised 1959, Methuen, 44.

[21] C. G. Seligman and B. Z. Seligman, *The Veddas,* Cambridge, 1911, Cambridge University Press, 106ff.

[22] V. Reynolds, 'Kinship and the Family in Monkeys, Apes and Man', in *Man* (new series), 3, 1968, No. 2, 209-23.

[23] For discussion along this line see M. R. A. Chance 'The Nature and Special Features of the Instinctive Social Bond of Primates', in *Social Life of Early Man,* 17-33 (see reference 15).

[24] V. Reynolds has elaborated this idea—e.g. 'Open Groups in Hominid Evolution', *Man* (new series), **1**, 1966, No. 4, 441-52.

[25] C. M. Turnbull, *The Forest People,* London, 1961, Chatto and Windus, 102.

[26] E. M. Weyer, *The Eskimos,* Hamden, Connecticut, 1962, Archer Books, 209.

[27] H. E. Driver, *Indians of North America,* Chicago, 1961, University of Chicago Press, 402.

[28] T. E. Rowell, R. A. Hinde and Y. Spencer-Booth, '"Aunt"-Infant Interaction in Captive Rhesus Monkeys', *J.Anim.Behav.,* **12**, 1964, 219-26.

[29] J. van Lawick-Goodall, 'Mother-Offspring Relations in Free-ranging Chimpanzees', in *Primate Ethology,* D. Morris (ed.), London, 1967, Weidenfeld and Nicholson, 308n.

[30] W. A. Mason, 'The Social Development of Monkeys and Apes', in *Primate Behavior,* 538 (see reference 3).

[31] J. P. Scott and J. L. Fuller, *Genetics and Social Behavior of the Dog,* Chicago, 1965, University of Chicago Press, 110-16, 147-50.

[32] For discussion of play in primates see C. Loizos, 'Play Behaviour in Higher Primates: A Review' in *Primate Ethology,* D. Morris (ed.), London, 1967, Weidenfeld and Nicholson, 176-218.

[33] H. H. Harlow and M. K. Harlow, 'Social Deprivation in Monkeys', *Scient.Am.,* **207**, 1962, 136-46.

[34] S. A. Queen, R. W. Habenstein and J. B. Adams, *The Family in Various Cultures,* Chicago, 1961, J. B. Lippincott, 134.

[35] E. E. Irvine, 'Children in Kibbutzim: Thirteen Years After', *J.Child Psychol. Psychiat.,* **7**, 1966, 167-78.

[36] For a discussion of the role of basic physiological and psychological factors which may underlie the incest taboo see R. J. Fox, 'Sibling Incest', *Brit.J.Sociol.,*

13, 1962, 128-50 and A. P. Wolf, 'Childhood Association, Sexual Attraction and the Incest Taboo: A Chinese Case', *Am.Anthrop.*, **68**, 1966, 883-97. Also see A. B. Reynolds, 'Kinship and the Family in Monkeys, Apes and Man', *Man* (new series), **3**, 1968, No. 2, 209-23, where the point is made that mother-son matings seem to be avoided in Japanese macaque monkey societies. D. F. Aberle *et al.*, 'The Incest Taboo and the Mating Patterns of Animals', *Am.Anthrop.*, 65, 1963, 253-65, develops the idea that the incest taboo was adaptive because it limited the bad effects of very close inbreeding. For a general discussion of the problem of incest taboos—both their genesis and persistence—see R. Fox, *Kinship and Marriage*, Harmondsworth, Middlesex, 1967, Penguin Books, Chapter 2.

[37] B. Kaminer and W. P. W. Lutz, 'Blood Pressure in Bushmen of the Kalahari Desert', *Circulation*, **22**, 1960, 289-95; J. L. J. Nye, 'Blood Pressure in the Australian Aboriginal with a Consideration of Possible Aetiological Factors in Hyperpiesia and its Relation to Civilisation', *Med.J.Aust.*, **2**, 1937, 1000.

[38] R. Singer, 'Some Biological Aspects of the Bushmen', *Z.Morph. Anthrop.*, **51**, 1960, No. 1, 1-6.

[39] J. L. J. Nye, 'Blood Pressure in the Australian Aboriginal', *Med.J.Aust.*, **2**, 1937, 1000.

[40] F. L. Dunn, 'Epidemiological Factors: Health and Disease in Hunter-Gatherers', in *Man the Hunter*, I. DeVore and R. B. Lee (eds.), Chicago, 1968, Aldine, 221-8.

[41] J. F. Roney, Jr., 'Palaeoepidemiology: An Example from California,' in *Human Palaeopathology*, S. Jarcho (ed.), New Haven and London, 1966, Yale University Press, 100.

[42] C. Wells, *Bones, Bodies and Disease*, London, 1964, Thames and Hudson, 54.

[43] M. Ch. Ehrstom, 'Medical Investigations in W. Greenland 1948/9', *Acta med. scand.*, **140**, 1951, 239-64.

[44] A. A. Abbie, 'Physical Changes in Australian Aborigines Consequent upon European Contact', *Oceania*, **31**, Dec. 1960, No. 2, 140-4.

[45] P. D. MacLean, 'Phylogenesis' in *Expression of the Emotions in Man*, P. H. Knapp (ed.), New York, 1963, International Universities Press, 26.

[46] A. Korlandt and M. Kooij, 'Protohominid Behaviour in Primates', in *The Primates*, Symposia of the Zool.Soc. of London, No. 10, London, 1963, 61-88.

[47] For an article suggesting that, in some cases at least, the victims were killed *in order* to eat the brains see A. C. Blanc, 'Some Evidence for the Ideologies of Early Man', in *The Social Life of Early Man*, 119-36 (see reference 15 above) and for the view that the skulls may merely represent the remains of funerary rites see F. M. Bergounioux, 'Notes on the Mentality of Primitive Man', 106-18 in the same volume.

[48] R. F. Heizer, 'The Western Coast of North America', in *Prehistoric Man in the New World*, J. D. Jennings and E. Norbech (eds.), Chicago, 1964, University of Chicago Press, 126-9.

[49] G. Silberbauer, *Bushman Survey*, Gaberones, 1965, Bechuanaland Govt, 78-81.

[50] R. R. Sears, E. E. Macoby, and H. Lewis, *Patterns of Child Rearing*, New York, 1957, Harper and Row, 266.

[51] G. Gwinnup *et al.*, 'Effect of Nibbling Versus Gorging on Glucose Tolerance', *Lancet*, **2**, 1963, 165-7.

Comments

L. R. HIATT A lot of civilised people would take exception to Dr Barnes's opening statement, that civilisation dates from the neolithic revolution. Quite properly, however, she has not allowed ethnic vanity to dictate the meaning of the title of this symposium, which on her definition of 'civilisation' now reads something like: 'The impact of settled agricultural and industrial society on the biology of man'. The contrast thus proposed is not industrial versus pre-industrial man, but sedentary agriculturalists and industrialists versus nomadic hunters and gatherers.

There can hardly be any doubt that two regular and prominent features of civilised society are territoriality and hierarchy. Yet Dr Barnes has suggested that neither appears to have been particularly important in palaeolithic society. It might be useful if I spend my time discussing some implications of this assertion for biologically-based theories of territoriality and hierarchy in man.

We can distinguish three views about human territoriality. The first, expressed by Carpenter in 1940 and later supported at length by Ardrey, is that 'possession and defence of a territory which is found so widely among the vertebrates, including both the human and sub-human primates, may be a fundamental biologic need'.[1] The second view, advanced in 1967 by Morris, is that a million or so years ago man emerged as a territorial hunter from non-territorial primate stock. The third view, proposed today by Dr Barnes, is that territoriality has become a pronounced feature of human social life only over the last 10,000 years or so and is to be seen as a by-product of a settled agricultural and industrial economy. Let me spend a moment indicating the developing empirical basis for this sequence of views.

Carpenter carried out pioneer field observations of monkeys and apes and was certainly correct in saying that territoriality is widespread among vertebrates. But later studies, particularly of gorillas, chimpanzees, and even baboons, have made it clear that not all primates are territorial in the sense of possessing and defending a delimited area. Indeed, Morris has asserted flatly that primates are not typically territorial animals.[2] The exception, in his view, is man the carnivore who, unlike some of his meandering, frugivorous cousins back in the forest, hunted in male packs, defended group territories against aggressive neighbours and, within the hunting domain, established individual home bases for pair-bonded breeding units. Change a word here and there and, as Morris sees it, this is pretty much the situation in the Western world today. Man acquired territoriality along with carnivorousness, and both have persisted as parts of his biological heritage.

An obstacle to Morris's view is the growing body of data indicating that contemporary hunter-gatherers have flexible, open territorial arrangements comparable with those of forest-dwelling primates. There seem to be good ecological reasons why groups of hunter-gatherers should behave hospitably to each other or avoid each other, rather than fight over each other's territory. And indeed reports in the ethnographic literature of territorial struggles between hunter-gatherer groups are extremely rare. We might infer, therefore, that palaeolithic man was, from the point of view of territoriality, a typical rather than an exceptional primate. The question would then arise in what sense territoriality as manifested in civilised man is to be understood at all as a biological phenomenon.

Dominance relations are a regular feature of primate life, though it is perhaps only in the ground-living baboons and macaques that hierarchy is a central organising principle. Ardrey and recently Morris both account for human dominance orders in phylogenetic terms. But again hunter-gatherer society appears to be a stumbling-block. Typically it is egalitarian rather than hierarchic. Among Australian Aborigines, for instance, there is a deeply-felt ideology that one man is as good as any other, and European administrators have regularly been frustrated in their attempts to develop a hierarchy within the local community. Morris suggests that dominance strivings may have become muted in early human society by the necessity for co-operation within the male hunting band, but before such a view could be accepted it would have to be demonstrated that palaeolithic man did characteristically hunt in organised bands. As often as not among contemporary hunter-gatherers, men hunt individually. Moreover, it is not clear in what respect an egalitarian band would be a more efficient hunting unit than a hierarchical one.

I have used the word 'egalitarian' to refer to hunter-gatherer society, but it is not a good term and should not be taken literally. Women and children, for instance, are subordinate to men, and commonly mature men are superordinate to youths. But formal hierarchies of individual males are a feature of sedentary communities, not of nomadic bands fluctuating in size and membership.

Dr Barnes has reminded us that, for at least 99 per cent of man's history, he has been a hunter-gatherer. This means that for the bulk of his time he has been a creature with plenty of elbow room, in two respects. First, on the great expanding frontier of human population it was possible for individuals to pack up and move on, though what kinds of pressure prompted what kinds of individuals to forsake their home range and begin afresh remains a matter for speculation. Second, even within a home range it was probably common for people to vary their associations considerably and to move about freely. This was certainly the case among Australian Aborigines. And here I should say in passing that the alleged 'walkabout' instinct is, in my view, largely a desire to slip away, temporarily at least, from the weight of white authority.

It should be clear that I support Dr Barnes's statements about the relative unimportance of territoriality and hierarchy in palaeolithic man. I have no means of judging, however, to what extent territorial and dominance behaviour in civilised man are culturally-conditioned or to what extent they represent a reactivation of long-dormant biological forces. Some years ago my colleague, Derek Freeman, remarked to me that the study of monkeys and apes in artificially-confined zoo conditions is in some respects of more relevance to an understanding of man than are studies made in the wild. Man, in other words, has more in common with caged apes than wild ones. An important implication of Dr Barnes's paper, it seems to me, is that the period of human captivity began some 10,000 years ago when man shackled himself with crops and herds.

1 R. Ardrey, *African Genesis*, London, 1961, Collins, 37.

2 D. Morris, *The Naked Ape*, London, 1967, Jonathan Cape, 30.

RICHARD E. BARWICK Dr Barnes has presented us with a very comprehensive and complete assessment of the major features of the biology of palaeolithic man. This symposium, concerned as it is with the effects of settled and agricultural society upon the biology of man, and our ability to assess the impact of civilisation, rests upon our recognition and understanding of the basic biological properties of man. Dr Barnes has very properly emphasised the extremely long period of man's palaeolithic existence, compared with the brief explosive manifestation of agricultural and industrial man.

From the zoologist's point of view, in man's lack of biological specialisation lie many clues to his biological success if we measure this in terms of 'the standing crop' or 'biomass' of the ecologist. His comparatively large size, as mammals go, removed him as a potential prey item for all carnivorous birds, reptiles, and mammals with the exception of the big cats and dogs hunting in packs. It also allowed him to utilise, without restrictive specialisation, the entire range of food size, and through the use of simple tools to extend the range of items to include otherwise unobtainable plant and animal foods. Thus we have an animal with an enormously broadened scale of food size and with a diet that includes a surprisingly large proportion of the total food resources of the terrestrial environment.

As most of us would agree, arguments of the 'nature-nurture' variety should be avoided here. The biological basis of many human traits extends well into the past. Beach[1] has pointed out that there is a progressive emancipation of sexuality from hormonal control running through the primate order, and this appears to be paralleled by a progressive development from promiscuous mating to the formation of relatively long-lasting heterosexual partnerships between specific animals. The persistent interplay of traits inherited from man's subhuman primate past and the increased flexibility of his behaviour provided by his human brain has been properly stressed by Dr Barnes throughout her paper.

Some anthropological studies have attempted to relate human behavioural patterns to biological adaptation, and to explain the distribution of particular traits in human populations on the basis of their adaptive significance. One can briefly instance post-partum sexual taboos and late weaning of infants commonly occurring in tropical areas, and the early weaning and short post-partum sexual taboos common to temperate areas. Whiting[2] has suggested a correlation of these behavioural patterns with tropical low-protein diet, and the protection against protein malnutrition offered to infants by late weaning. Similar arguments have been advanced regarding the eating of wild or domestic animals which have died from natural causes. Societies either taboo such foods or permit their use. The gain in terms of dietary protein must be balanced against possible harmful effects from disease transmitted from animal to man.

For the zoologist, man's relationship with other animals is of interest. Dr Barnes has not mentioned the domestication of other animals, which we usually associate with the neolithic peoples, as the first step to animal husbandry. Whether it occurred once or several times, the domestication of the dog must have taken place early in the sequence. Thus the hunter-gatherer adopted a powerful ally in the hunt (many modern breeds whose origins are known were evolved for hunting purposes). At the same time, the dog is the one other terrestrial mammal which hunts co-operatively, and one which competed directly with the human hunter.

Bartholomew and Birdsell[3] have stressed two major ecological points with regard to pre-agricultural man. Firstly, that 'the basic problem of human behaviour, like the behaviour of other animals, is the obtaining of food, for the human body requires a continuous input of energy, both for maintenance and propagation'. The second point relates to one of the prime concerns of the ecologist—the size, structure, and density of human pre-agricultural populations. The population density of animals is normally in a complexly maintained equilibrium, dependent in the case of man upon environmental, behavioural, and cultural forces. All animal populations depend upon the 1 per cent of incident solar energy which plants are able to capture, and there is continuous competition among them for this 1 per cent. Individual competitive success can be measured in terms of metabolism and the success of a population as the product of population density times individual metabolism. Comparatively few attempts have been made to measure the competitive success of hunter-gatherer societies. Pearson[4] provided data for a comparison of Indians of the north-eastern United States with other animals of the area. Indians had less metabolic impact than deer, and about the same impact as long-tailed shrews. Whatever the absolute validity of such figures they do indicate that the environmental impact in metabolic terms of pre-agricultural man was at least in the same order as that of other animals.

Food-gatherer and hunter had access to the widest range of environments. The populations were never large, but as Deevey[5] has estimated the upper palaeolithic and mesolithic populations of man totalled about 5 million, an average of ·04 persons per sq kilometre (·1 person per sq mile of land). With the agricultural revolution, population moved up two orders of magnitude to a new plateau; to an average of 1 person per sq kilometre, 8,000 years later. The increase over the last 300 years is a sixfold one, and now there is an excess of 20·5 persons per sq kilometre.

Deevey also estimates that there were 36 billion palaeolithic hunters and gatherers including the first tool-using hominids; another 30 billion may have walked the earth before the invention of agriculture. A cumulative total of about 110 billion individuals have passed their days on the earth. As he comments:

> 'Neither for our understanding of culture nor in terms of man's impact upon the land is it a negligible consideration that the patch of ground allotted to every person now alive may have been the lifetime habitat of 40 predecessors.'

Biological adaptability notwithstanding, cultural adaptability often led to technological solutions which were perfectly fitted to the societies' needs in particular environmental situations. As an example, the birch-bark canoes of the Cree Indians, hunters and gatherers of northern Canada, were the most efficient transport solution in this well watered country, but such an invention did not have the technological potential of the wheel, and was a sort of cultural dead end. And again, such inventions as the wheel may not be effectively utilised; the wheel was independently invented by the Indians of Middle America, but was used only on children's toys, for these irrigation communities already had a slave labour force which was effectively used to carry burdens on their roads and there were no native mammals suitable for domestication as draft animals.

The paradox is that the rapidity of cultural change keeps behavioural outcomes continuously off balance. These outcomes are unbalanced further by radical changes in the environment, themselves brought about by the results of human behaviour.

1 F. A. Beach, *Sex and Behavior*, New York, 1965, Wiley.

2 J. Whiting, 'Effects of Climate on Certain Cultural Practices' in *Explorations in Cultural Anthropology: Essays in Honor of G. P. Murdock*, W. Goodenough (ed.), New York, 1964, McGraw-Hill, 511-44.

3 G. A. Bartholomew and J. B. Birdsell, 'Ecology and the Protohominids', *American Anthropologist*, **55**, 1953, 481-98.

4 O. P. Pearson, 'Metabolism and Bio-energetics', *Sci. Monthly*, **66**, 1948, 131-4.

5 E. S. Deevey, 'The Human Population', *Scient.Am.*, **203**, 1960, 194-204.

R. STORY I intend speaking on some controversial things, on some material that Dr Barnes probably has not had access to, and on some of the things in her paper, and shall begin by saying a little more on what she has said about the nuts used by the Bushmen in South West Africa. The tree in question is very likely to be *Ricinodendron*; sturdy, squat, and branching from the base. Where the stems join at the bottom, they have the habit of rotting away to form an underground hollow that is a natural reservoir for rain water, which trickles down from the branches and stems during the rainy season. For the Bushmen this ensures a supply of food and drink of a quality that we ourselves, well fed as we are, would not be scornful of. But although the Bushmen have a sound argument when they say 'Why should we plant when there are so many of these nuts around?', they are unsound in respect of something equally vital. In contrast to the food supply, the water supply is limited—the reservoirs operate at about half capacity because of wind-blown sand, the result being that the Bushmen are out of water long before the nuts are exhausted. The difficulty could be corrected by a few minutes' communal work in scooping out the sand. That it is not, indicates the idleness and improvidence that are familiar in our own society. It may indicate also that the Bushman attitude towards planting could stem from the same two causes rather than from wisdom.

Dr Barnes has quoted someone's theory that the switch from gathering to hunting would have been accompanied by an increase in the efficiency of hand/eye co-ordination and visual acuity and hearing. However, it is an education in the use of sharp vision to go out with a band of foraging Bushmen. They miss nothing, including underground vegetables betrayed by cracks in the ground or by tiny and inconspicuous scraps of tissue on the surface. In view of the Bushman skill in foraging, it seems doubtful if we can correlate hunting with increased visual acuity, because for the gatherer also it is perhaps equally important to have very good eyesight indeed. It would on the other hand be reasonable to suppose that the sense of smell would increase with the switch from gathering to hunting, for it is a tremendous advantage for a hunting animal to perceive the hunted without in turn being perceived. Yet we, who have gone through the hunting phase, or rather are still in it, have a notoriously poor sense of smell. Colour vision may not always be a help. It sometimes happens, when one is working with aerial photography and using colour and black and white film for the same area, that the contrast in black and white is a good deal better than in colour, where there are not two main subdivisions but a large variety, in which slight gradations are liable to confuse one. To sum up, it may be a rather hasty conclusion that an increase in the complexity of living necessarily gives rise to an increase in perceptivity.

I may add a little too to the question of the extrovert and his use of nicotine. If one offers a cigarette to monkeys or baboons in captivity, one will find that they will peel the paper off and eat the tobacco; and as far as psychological make-up is concerned, the monkeys' cage is the place to go if one is looking for extroverts. The Bushmen have a similar craving. Those in the more remote parts are not habitual users, for they see tobacco only in their rare contacts with the Bantu. Yet they are extravagantly eager to obtain it, and immoderate in its use when they do, to the point of being rendered unconscious by the continual inhaling. There are evidently good grounds for the suspicion that has been expressed of a very deep-seated and primitive need for something contained in tobacco, whether lighted or not.

Although I never saw the Bushmen involved in any fights, Dr Barnes is still not quite correct in saying that they are called 'the harmless people'. That is what they call themselves, which may be a very different matter, and in fact there are references to violence among the Bushmen in Bleek's *Bushman Folklore*. As to hazards, the Bushmen we were in contact with were in a remote area of big game in its natural state, but there was a kind of armed neutrality between them and the predators. Except very rarely, the lions and the Bushmen do not molest one another, and even small children wander freely with little danger except from snakes. The hazards come more from the lack of medical attention, otherwise I do not think they are higher than those we encounter in our modern form of life.

Discussion

BARNES Yes, I agree that the Kalahari Bushmen call themselves 'the harmless people'—but I think that observation of them has indicated that this is a fairly reasonable epithet worthy of being perpetuated in the literature.

Turning to the problems of visual acuity and sense of smell, I would agree that gathering in a harsh environment, such as the Kalahari desert, requires good long distance vision. It is an advantage for women as well as for men—and could be selected for in both sexes. The comparison I was thinking of was with the forest dwelling fruit, leaf, and pith eaters such as the chimpanzee. Investigations in laboratory animals of this species have shown a tendency to myopia. As for smell—most carnivorous mammals do hunt by smell, but I think that when our ancestors took to hunting they were already so far advanced along the primate pathway where vision is the dominant sense and smell a subordinate sense, that they became this mammalian oddity—a carnivore hunting by sight not smell![1]

1 B. G. Campbell, *Human Evolution*, Chicago, 1966, Aldine, 294.

MIMS I was interested when Dr Barnes referred to the taboos on incest in man. As far as I know there is no such thing in animals, but this could be simply because they have not been studied adequately. For the taboo to work, of course, one has to be able to recognise one's own kin at sexual maturity.

MCBRIDE I would like to comment on this question of the incest taboo, which is often alleged not to occur in animals. The point is that not many species retain their offspring after sexual maturity, and therefore the question of incest taboo rarely arises. Those that do retain their offspring until a stage when they reach sexual maturity, usually have some mechanism for inhibiting its onset. The kookaburra retains its offspring, and so also, it is alleged, does the wolf; and in these cases there is an organically controlled mechanism to limit incest by the suppression of sexual maturity by the parents.

I would like to make another comment about this use of the word 'territoriality'. It is a hopeless word, and it is bandied around in all sorts of situations. I don't like it. But it does describe one category of spacing phenomena. The important issue seems to be that animals do space themselves. They space themselves characteristically in a number of ways. Spacing merely says that animals maintain areas around them free of conspecifics of one form or another, perhaps just males or perhaps other females. Now if animals do this, they can do so by attaching to fixed space and keeping other individuals off the fixed space. This is an additional part of the spacing mechanism. When this occurs we have one of the very many forms that we can observe of territoriality. Territoriality is not a phenomenon, it is a whole class of behaviours. But animals space in other ways. They may move over home ranges, perhaps remain solitary—this means that there are mechanisms for preventing them from coming into contact with one another. This one can observe if one watches home ranging animals when they do, in fact, become close. They defend a portable area, a portable spacing area as they move around a home range, so remaining in isolation.

Now, once we say that animals control a certain amount of space we immediately raise the point of dominance, because dominance or submission cannot be considered without some space. An animal is not a stimulus to aggression if it is at a certain distance or on a certain piece of land. If another animal comes within that area then dominance would be expressed, and submission on a territorial system is merely flight to the boundary of that territory, or in the home ranging system it is flight to the distance of the personal sphere. We do not know whether the home ranging animals are open systems, we do not know whether if you poke a new animal from fifty miles away into this system it will merely be able to move around on a home ranging system. This is particularly relevant in the animals like the chimpanzees which are home ranging small groups of males and females which occasionally aggregate and disperse and form other aggregates and disperse. We accept that these are open among the animals that they will normally meet on a common home range—but we do not know what would happen if we moved another animal from fifty miles away into this area. Now, even in groups there is spacing behaviour, and this again is an expression of dominance. An animal cannot move into the immediate vicinity of a dominant individual but a dominant individual can control the movement of a subordinate within the group, by merely getting it to avoid or submit by moving away.

FREEMAN I happen to have with me a recently published paper by Dr Donald S. Sade, of Northwestern University.[1] It reports the discovery of an inhibition of mating between son and mother in a species of non-human primate (*Macaca mulatta*), this inhibition being associated with the dominance of the mother over her male offspring in the course of ontogeny. This discovery has, it is plain, important implications for hypotheses as to the origin of incest prohibitions in human societies.

[1] D. S. Sade, 'Inhibition of Son-Mother Mating among Free-ranging Rhesus Monkeys' in *Science and Psychoanalysis*, **12**, New York, 1968, Grune and Stratton, 18-38.

J. M. Rendel

The Time Scale of Genetic Change

The most characteristic differences between men and other primates are associated with the increase in human brain size and the development of mental abilities that go with this increase. It is appropriate therefore to show the rate at which the human species has been evolving in terms of changes in cranial capacity and the appearance of new skills. Table 2:1, which is taken mostly from Dobzhansky's book *Mankind Evolving*,[1] shows in outline how much time has elapsed between different stages in human evolution. It is set out in years and in generations. The number of generations is calculated on the assumption that the generation length 1,000,000 years ago was about 20 years and rose slowly to perhaps 35 years by the nineteenth century and is now slipping back again as the age at which women have their children is reduced. The average age of the mothers of children born in Finland in a given calendar year was just under 27 years in 1965, in 1890 it was 30·9 and in 1878 it was 33·7.

Table 2:1

Generation intervals between the appearance of different human skills

Years prior	Species found	Skills	Cranial capacity cc	Generations prior to 1965
1,000,000	Australopithecus	Pebble tools	450-500	50,000
500,000	*Homo erectus*			
	Java	Tools	770-1000	25,000
	Peking	Tools, fire	900-1200	
75,000	*Homo sapiens*			
	Neanderthal	Tools, fire, skins, burial	1300-1425	2,500
40,000	Sapiens	Tools, fire, skins, burial, arts	1200-1500	1,300

About 50,000 generations ago there lived an animal, Australopithecus, not quite human according to some authorities, with a brain which was small by modern standards; he walked upright and made simple tools. Twenty-five thousand generations later, an animal could be found whose brain size was double that of Australopithecus and who was making simple tools and using fire. After a further 22,500 generations human brain size was almost treble that of Australopithecus, and the preparation of skins and burial of the dead had been added to the list of his accomplishments. Twelve hundred generations later, that is 1,300 generations ago, modern man had appeared without further change in brain size but with the addition of the practice of art to his list of accomplishments. From then on, civilisation has emerged at lightning speed, but I think there is no reason to believe that there has been any acceleration in man's physical evolution or indeed any great change in his intellectual capacity. Going back two or three thousand years we find people like Archimedes, who would have been a first-class scientist today, and Pythagoras, who must be rated by modern standards as a first-rate mathematician, and the poets and dramatists of those times who are still read and performed today. Looking back from these people of 100 generations ago whom we know, to those we do not know who invented metallurgy, the wheel, numbers and the alphabet, who built Stonehenge, tamed animals, domesticated plants, and learnt to brew, and even to those who learnt to use fire in the days before modern man, we cannot suppose the human species has changed much in innate abilities over the past 500 generations and only a little, if at all, over the whole 1,300 generations of modern man's known existence. Though his cultural evolution has been explosively fast over the past 400 generations, all the evidence suggests that the pace of his organic evolution is much what it was 50,000 generations ago and certainly no faster. It might even have slowed down a little to approach the rate of advance in the horse which has taken some 3,000,000 generations to get from Merichypus to where it is now.

There are nevertheless different sorts of changes which may be going on, some on quite a short time scale. The following are the sorts of changes which may be taking place:

1. Slow long-term trends which confer on individuals new properties of advantage in most or all environments. The increase in brain size is an example. The increase referred to is that of the cerebral hemisphere, which is associated with increasing mental ability in primates, and is correlated with increase in cranial capacity.

2. Shorter-term trends due to strong selection pressures of relatively short duration brought about by changing environmental conditions such as the appearance of a new disease, the discovery of a new way of life, or the introduction of a new component to the diet.

3. Trends due to changes in selection pressures which upset preexisting equilibria. Gene frequencies that are no longer at equilibrium levels will move towards the new equilibrium level

(a) by mutation alone if the genetic variations in the character concerned are neutral

(b) by selection if there is a selection differential.

4. Immigration and mixing of populations. This is undoubtedly

important but can be regarded as an upheaval which resets a starting point from which the other three take off.

There is little to be said about the slow long-term trends. If we take brain size as an example, we can guess at the sort of pressures at work from the changes that have taken place by making a couple of assumptions. In 50,000 generations cranial capacity has gone from 475cc to 1,400cc. Suppose that increase is proportional to size, that is to say, the coefficient of variation is roughly constant and increases are geometric; suppose that about half the variation in cranial capacity is genetic, that is, heritability is 50 per cent. The increment each generation is by an amount a, so $475 \times a^{50,000} = 1400$ and $a = 1 \cdot 00002$, so selection differential must have been of the order of ·00004 each generation. This means that the mean cranial capacity of people who became parents of one generation, weighted for the number of children they had, must have been about ·019 cc larger than the average of the whole population at the beginning of the 50,000 generation period and about ·056 cc in the recent past. In a long-term trend of this sort a very small selection differential is at work. It is important to note that a vitally important evolutionary change such as increase in cranial capacity, which by evolutionary standards has been a fast change, would be utterly undetectable to us if it were going on today. The past 1,300 generations might be expected to have added 40 cc to cranial capacity.

This can be contrasted with the sudden exposure of a population to a strong selection pressure like the arrival of the plague in Europe. This disease appears to have entered Europe in about 600 A.D. and to have left again in 1700 A.D. We can take it that the European population was fairly thoroughly, though not universally, exposed for about thirty-five generations. If we make the assumption that the ability of Europeans to survive the plague has been built up by selection and that a population exposed for the first time will be about as susceptible as the Europeans of 600 A.D., we can make a guess at the intensity of selection the plague imposed and use it as a guide to the sort of selection pressure on a human population which can be considered fairly intense. Case mortalities all over the world lead one to expect 35 per cent of Europeans infected to die of plague. On the other hand, case mortality in Hong Kong was over 90 per cent and in Bombay over 75 per cent where the disease struck populations never previously selected for resistance to plague. Suppose that thirty-five generations of selection has raised survival rate from 25 to 65 per cent. This is an increase of one standard deviation, or $\cdot 03\sigma$ each generation. We have no guide to heritability in this case; by analogy with myxomatosis in rabbits it might be of the order of 30 per cent; if so, resistance of parents weighted for the number of their offspring would have had to be $0 \cdot 1\sigma$ better than the resistance of the population from which they were drawn. This compares with a change in brain size of ·002 per cent in each generation in response to a selection differential of ·004 per cent. The selection against people susceptible to plague is very much more intense. Even so thirty-five generations have not moved the mean resistance of the population by more than 1σ. Thus a thousand years is the sort of time it will take to make a change equal to one standard deviation in a human trait when selection is fairly intense. The standard deviation of I.Q. is about 15 units and it is extremely doubtful whether selection is strongly in favour of or against intelligence at the present time. It follows that in a thousand years' time intelligence as measured by I.Q. will be within 2 or 3 units up or down of what it is now, unless something drastic happens to alter the current position.

This brings us to consideration of gene equilibria and the ways they may be established. Civilisation must have imposed selective forces on the population not present before and must have relieved the population from others which were maintaining gene frequencies at a low equilibrium level, thus upsetting the equilibrium between mutation and selection which becomes established in a stable gene pool. Once a genotype has been established in a population as the best in the circumstances, mutation from it to anything else is almost certain to be bad. So long as the selection pressures which established the favoured genotype continue, deviations from it will be kept rare. They will either die before they breed, or fail by living shorter lives and having fewer offspring than the favoured types. So mutation is continually feeding into the population unsatisfactory types which selection then weeds out. The balance point is reached when the number of new genes added to the gene pool by mutation in each generation equals the number eliminated by death and failure of those that carry these genes. This death and failure has been referred to as genetic load. So long as there is mutation, genetic load is inescapable. I have summarised the more drastic forms taken by genetic load in Table 2:2.

Table 2:2

Approximate number and type of loci responsible for genetic load

Causative genetic agent	Number of loci (approx)	Number of births % of all births
Single autosomal recessive	30	0·17
Single recessive sex linked	8	0·04
Single autosomal dominant	115	0·60
Genetic possibly polygenic	?	1·66
Genetic susceptibility possibly polygenic	?	0·98

The summary is taken from a report to the United Nations[2] and is based on statistics collected in Northern Ireland. I have taken several liberties with these. I have left out all conditions which I consider too mild to be a serious handicap. This reduces the incidence of autosomal dominant abnormalities from 0·9555 to 0·60. It also reduces the incidence of autosomal recessive conditions a little but this reduction is more than made up for by including cystic fibrosis which the

UN report omits but I include. I shall discuss this condition in detail later. In addition, I have made an increase to the number of gene loci supposed to be involved. There are forty-nine recognisably different dominant conditions and this is the minimum number of loci responsible; but many syndromes studied by geneticists in organisms in which the appropriate tests can be made have proved to be arrived at by more than one route and if the mutation rate, assuming one locus only, seems very large, I am assuming that we are really dealing with two or more separate loci, each with the same phenotype. For example, most dominants which appear to be single genes and whose mutation rate has been estimated, mutate from normal to the dominant form at the rate of $2 \cdot 5 \times 10^{-5}$. Mutation rates much greater than this are suspicious. Hydrocephaly of the internal obstructive type, which is quite lethal, has an incidence of 1·2 in 1,000 births. If a single gene were entirely responsible, the mutation rate to it would have to be of the order of 1 in 1,000 which is forty times the average rate. There are, no doubt, a number of reasons why the condition is so frequent. One possibility is that several genes lead to the same end result so that though, if we were to make assumptions about penetrance and fitness, the overall mutation rate might be as high as 1 in 5,000, if we made the further assumption that there are 4 loci, the mutation rate at each locus would only be 1 in 20,000. This is admittedly quite arbitrary. I have assumed the forty-nine conditions are the result of mutation at 115 loci; I may be wrong; the number could not be less than forty-nine. One hundred and fifteen loci would have to mutate at the average rate of 1/40,000 to be in equilibrium with an average fitness of 0·25. The frequency of living people with the forty-nine conditions I have included is 3·60 per thousand instead of 6·0 per thousand as at birth, so allowing for a drop in fertility as well as the effect of this differential survival rate, 0·25 is not very far out. The assumption of 30 loci for the twenty-six recessive conditions listed in the UN report is really more or less a rounding off operation. No individual recessive condition, with the exception of cystic fibrosis, was very frequent. I have taken the sex linked conditions as they stand. There remains a majority of conditions with a genetic background that is not yet fully understood. Some of the conditions listed may turn out to originate from mutation in a single autosomal gene, others to be polygenic; in either event I shall leave these two classes of conditions on one side.

Let us first consider what factors determine the frequency of conditions brought about by the mutation of single major genes. There are two opposing forces, mutation and selection, whose balance determines the equilibrium gene frequency. When the heterozygote has an advantage over both homozygotes we have a special kind of equilibrium. Equilibrium is upset when selection forces or mutation rate change. The new equilibrium is achieved at a rate dependent on selection and mutation.

In the absence of selection the only forces to be considered are those of opposing mutation. If we have a gene A mutating to a recessive form a at the rate of β in each generation and a mutating back to A at the rate of γ, equilibrium will be reached when the number of new As exactly equals the number of new *a*s; we can write this $A \underset{\gamma}{\overset{\beta}{\rightleftharpoons}} a$. If the frequency of A is $1 - q$ and *a* is q we can write $(1 - q)\,\beta = q\gamma$. In the absence of selection, the frequency of A and *a* depends on the relative mutation rates. The time taken to reach equilibrium from a point of disturbed equilibrium will also depend on mutation rate and on q and $1 - q$; the rate of change in q diminishing as q and $1 - q$ approach equilibrium values. The order of magnitude of mutation rates in human genes is $2{\cdot}5 \times 10^{-5}$, so rates of change are very slow when only mutation is involved. Before converting these rates into terms of years or generations, let us see how selection adds to the picture. Suppose *aa* individuals die more readily and reproduce less readily than AA or A*a* individuals and that for every one individual passed into the next generation by AA or A*a* parents, $1 - S$ are passed on by *aa* parents; at equilibrium the number of *a* genes lost from the population due to death and failure to breed of *aa* must equal the number replaced by mutation. The proportion of *a* genes lost is q^2S, in addition γq mutate to A. So $(1 - q)\,\beta = q\gamma + q^2S$. It is possible to simplify this for rare genes, such as lethals for which S is 1 and all other genes with an S of more than $0{\cdot}05$, for $(1 - q)$ is then sufficiently close to unity to be taken as such and q so small that $q\gamma$ can be ignored and $\beta = q^2S$ becomes an approximation quite close enough for our purposes. Now β is $2{\cdot}5 \times 10^{-5}$ or 1/40,000 so q and q^2 can be estimated for genes that are lethals, $S = 1$, to genes that are mildly handicapped, $S = 0{\cdot}05$, as in Table 2:3.

Table 2:3

Relationship between strength of selection and equilibrium gene frequency where the gene is a recessive

S	q^2	q
1	1/40,000	1/200
0·8	1/32,000	1/179
0·6	1/24,000	1/155
0·5	1/20,000	1/141
0·4	1/16,000	1/126
0·2	1/8,000	1/89
0·1	1/4,000	1/63
0·05	1/2,000	1/45

This table shows how powerful S is at fixing the equilibrium value of q. Even a very mildly handicapped person with $S = 0{\cdot}05$ will be rare, q being roughly 1 in 45 and q^2, the frequency of the individuals themselves being of the order of 1 in 2000. The value of q is roughly the proportion of heterozygotes; q has to be 1/10 before 1 per cent of the population are homozygous. To reach this value from $q = 1/200$ by mutation alone—ignoring back mutation and the slight shrinkage of $(1 - q)$ and assuming mutation increases q at the rate of $2{\cdot}5 \times 10^{-5}$ a generation—would take 38,000 generations, which is three quarters of the whole geological history to date of more or less human beings. So when civilisation makes it easier for poor recessive genotypes to survive and finds ways of reducing the unfitness of some recessive genetic conditions, such as phenylpyruvic idiocy, which can be alleviated by feeding diets free of phenylalenine, we do not need to fear that a rapid increase in the frequency of the gene will follow. As long as there is any handicap, equilibrium frequency remains low. Incidence would reach as much as 1 per cent in a million years only when fitness really had been raised to unity—that is to say, once all handicaps were successfully removed—and if this could be done the character could not be considered harmful or objectionable any more.

Our first conclusion must be that fears, which are often expressed, that harmful effects to the genotype of the species will follow remedial measures taken by medicine against genetic abnormalities are groundless. Not only must the character treated be rendered harmless before equilibrium gene frequencies change very far, but the rate at which equilibrium gene frequency is reached is very slow. Our type, let alone our species, has not lived long enough for such an effect to have been noticeable even had our remote ancestors been practising modern medicine 1,000,000 years ago. Of course, remedial action does mean work by society. The death and failure of the genetically unfit—the genetic load—is converted into a social load and if society cannot bear the load it will not and some genetic load will remain, keeping gene frequency low by a continuous death and failure of some fraction of the homozygotes. The only real danger to the species is that it will build up too big a social load which, though easily tolerable in normal times, renders the species brittle and susceptible to any sudden change which might disrupt the organisation and smooth functioning of society.

The table does not show the effect of mutation rate on the equilibrium value of q^2. The equilibrium value of q^2 is in fact exactly proportional to mutation rate. Any major increases in mutation rate would simultaneously raise the equilibrium frequency of all recessive conditions. It would take a long time to reach the new equilibrium by mutation, but as mutation itself would have increased, not so long as at the old rate. It would be easy by careless use of radiation to multiply mutation rate by tenfold. This would have a dramatic effect in due course. But its most dramatic effect would be on dominant mutations.

If A' is dominant to A and lethal in homozygous condition, every mutant is obvious in heterozygous condition and if fitness is $1 - S$, and there are 2q heterozygotes, Sq mutant genes are lost each generation, so to a first approximation $\mu = qS$. If S runs from completely lethal to mildly handicapped we have Table 2:4, where $\mu = 2{\cdot}5 \times 10^{-5}$:

Table 2:4

Relationship between strength of selection and equilibrium gene frequency where the gene is a dominant

S	q	Heterozygotes
1·0	1/40,000	1/20,000
0·8	1/32,000	1/16,000
0·6	1/24,000	1/12,000
0·5	1/20,000	1/10,000
0·4	1/16,000	1/8,000
0·2	1/8,000	1/4,000
0·1	1/4,000	1/2,000
0·05	1/2,000	1/1,000

Once more it is noticeable how powerful S is. With a fitness of 95 per cent equilibrium gene frequency is 1/2,000 and 1/1,000 people will suffer; to reach a frequency of 1/2,000 from the one in balance with S = 1, mutation would have to add mutant genes totalling $\frac{19}{40{,}000}$, and at the rate of $\frac{1}{40{,}000}$ a generation, assuming this could be kept up, it would take nineteen generations. What happens to dominants is very much nearer the surface; the equilibrium values of q are smaller and so mutation reaches equilibrium faster—even so twenty generations (or 600 years) is quite a long time. What is important is that if μ is doubled over the whole genotype it takes a very short time to reach the new equilibrium level; if S is on average ·5 it would take not much more than four generations to attain the new equilibrium; at this point the frequency of dominant abnormalities at birth would be 12 per 1,000 and not 6·0. So mutation rate is of immediate importance to us, absence of selection against deleterious genes a remote danger. It follows that care should be taken not to increase mutation rate, and effort expended on ameliorating the condition of the genetically handicapped.

An interesting feature of Table 2:2 is the frequency of heterozygotes. For recessive lethals it is 1/200; for mildly disadvantageous conditions it is 1/45 so the frequency of marriages between like heterozygotes will be very small. The United Nations report[2] from which I have taken my figures lists some thirty conditions as making up the 0·1 per cent of births abnormal because of a homozygous recessive gene in Northern Ireland. The total of marriages liable to produce abnormal people of this kind is about 1/250; the average is 1/7,500 for each locus. If all heterozygotes could be detected, couples of like heterozygosity could be warned of the risk of having children. If even a small fraction of heterozygotes refrained, or if a number could be persuaded to have only one or at most two, a pressure much greater than any pressure selection against homozygotes can generate would be imposed on each gene and its frequency would begin to drop to a much lower equilibrium frequency. If a large percentage could be persuaded either to refrain from having children or to marry someone else the actual incidence of abnormal births could be lowered too.

The chance of spotting heterozygotes is very good indeed. New techniques still have to be evolved and current ones applied but in the normal course of events it is only effort and money which stand in the way of identification of heterozygotes. I believe that the time will come when all children born will be screened for the presence in heterozygous state of whatever happen to be the thirty or forty important recessives and the sixty or eighty important dominants. Already many hospitals make a routine check on all infants to see if they are homozygous for phenylpyruvic idiocy. This is done because if homozygous phenylpyruvic idiots are put on to a diet free of phenylalanine soon enough, brain damage can be avoided, but if feeding of the diet is too much delayed, permanent brain damage is done and the diet is only partly successful. Once laboratories are set up for making this sort of check, it will require little extra work and few extra facilities to screen for the rest, particularly if the tests can be automated.

Dominants are immediately obvious and no doubt part of their unfitness is due to the unwillingness of people suffering dominant disabilities to pass these on to their children. Assuming an overall frequency of 0·6 in a hundred there are only 1·2 per cent of marriages affected even if all marry and it should not be hard to persuade a sufficient number that their condition is heritable and that they should not have children to keep S high and frequency low. Of course these people themselves know what the subjective handicap is and are in the best position to judge whether they should risk having children who will suffer as they suffer. It should only be necessary to tell them of the risk.

In addition to the balance between selection and mutation, equilibrium gene frequencies are set up when the heterozygote Aa is fitter than either homozygote. There are some good examples of this sort of equilibrium in man, the classic being the gene for thalassaemia, lethal in homozygous condition, but highly resistant to malaria in heterozygous condition and common in those populations in Europe in which malaria is—or was until recently—endemic. In general, if S is the fitness of one homozygote, the frequency of whose gene is q, and t the fitness of the other, the frequency of whose gene is $(1 - q)$, and the fitness of the heterozygote is 1

$$S = \frac{t(1 - q) + 2q - 1}{q} \quad \text{and when t is 0}$$

$$S = \frac{2q - 1}{q}$$

It is tempting to explain every excess of heterozygotes beyond what one would expect from the number of homozygotes as due to hybrid vigour. But there are at least two other explanations possible. The first is that mating is not at random. The expectation that heterozygotes will appear with a frequency $2q\ (1 - q)$ only holds for random mating. As by custom and law the human species is largely outbred, one can expect a small excess of heterozygotes. But more probably where the discrepancy is large, one may find that q is out of balance with S. When selection is suddenly imposed upon a recessive gene, it sets a new equilibrium frequency for the gene whose frequency will start to decline but will not reach equilibrium at once and may not have reached the new equilibrium point at the time of investigation. Although the homozygotes have suddenly become unfit, possibly lethal, their incidence will still be frequent—far too frequent to be in balance with anything but an enormous mutation rate—because the number of heterozygotes has not as yet fallen to the proper equilibrium level. I suggest, and this is speculation, that cystic fibrosis may be a case in point. Dr Naughton has suggested that cystic fibrosis may be due to a failure of the mechanism which pumps salts back out of certain secretions. He bases his suggestion on the fact that the sweat of patients is five times as high in salt as the sweat of normal people, and that a characteristic of patients is the rapid crystallising of mucus in the ducts of secreting glands. When a salivary duct is canulated in a patient the mucus from it rapidly turns cloudy, whereas the mucus from the duct of normal people does not. It does in normal people, however, if a few parts per million of zinc, making the total about 6 ppm, is added; as patients are not removing salts from their secretions, the zinc content may well be up along with other salts and may be up enough to turn the mucus cloudy. Now if zinc is implicated, it is possible that at some time during the last 200 years the gene for cystic fibrosis has become much more dangerous than it was. About 200 years ago the process of galvanising iron was invented. Galvanised iron is iron covered with a layer of zinc and is used in roofing and guttering, water tanks and water pipes, feeding troughs, fencing wire, watering cans and so on to an extent which may conceivably have increased the zinc content in food and water to the point where people's blood content is one or two parts per million more than it was 200 years ago. The frequency of cystic fibrosis which is quite lethal is $\cdot 0006$ and at a mutation rate of $2\cdot 5 \times 10^{-5}$ would be in equilibrium with a fitness of just over 95 per cent. This suggests the possibility that its lethality was only recently acquired and it is not yet at the new equilibrium frequency, in which case the use of zinc may be the basic cause. If this story is anything like correct, a judicious use of a chaelating agent might work wonders. In fact acetylcystine, which does help when inhaled, does chaelate zinc.

There is no reason to suppose that this is the only example of an environmental change turning a neutral gene into a lethal one. Indeed it is possible that kuru is one. Whenever a gene appears too frequent for a reasonable mutation rate the possibility that some new environmental agent has recently made it lethal should be looked into.

Tables 2:5 and 2:6 show how fast single genes respond to selection of different strengths at different frequencies. The important point is that dominants respond fast even at quite low frequencies. But at

Table 2:5

Rate of change in gene frequency at different selection intensities and at different gene frequencies (recessive)

q	S				
	1·2	1·4	1·6	1·8	2
·5	·05	·1	·15	·20	·25
·1	·002	·004	·006	·008	·01
·05	·0005	·001	·0015	·002	·0025
·02	·00008	·00016	·00024	·00032	·0004
·005	·000005	·00001	·000015	·00002	·000025

Table 2:6

Rate of change in gene frequency at different selection intensities and at different gene frequencies (dominant)

q	S				
	1·2	1·4	1·6	1·8	2
·5	·09	·17	·23	·29	·33
·1	·018	·028	·057	·074	·091
·05	·0099	·0197	·0291	·0384	·0476
·02	·004	·008	·0119	·0158	·096
·005	·001	·002	·003	·004	·005
·000025	·000005	·00001	·000015	·00002	·000025

very low frequency even dominants increase slowly. A dominant with a mutation rate of $2\cdot5 \times 10^{-5}$ would have a frequency of ·000025 and increase at the rate of ·000025 a generation if it suddenly became twice as fit as the normal gene; this is very nearly doubling the gene frequency each generation, at which rate of increase frequency would reach nearly ·05 in twelve generations. A recessive lethal which acquired a fitness of 2 would start with a gene frequency of ·005 and increase it initially at the rate of ·000025 a generation. Increase in frequency of a recessive is proportional to the square of frequency and it would take more than twelve generations to increase frequency from ·005 to ·0054. So selection of a powerful kind will increase gene frequency fast if the gene is dominant at all frequencies, but only slowly if the gene is a recessive unless it is reasonably frequent already, say between ·05 and 0·1. But even with selection operating in conditions leading to rapid change, we are concerned with change of gene frequency of 1 or 2 per cent in centuries.

The most dangerous possibility from the eugenists' point of view is that a hereditary condition which is socially unwanted, and kept in check by selection because biologically disadvantageous as well, may lose its biological disadvantage and become biologically advantageous without losing its social disadvantages. Intelligence has been claimed to be such a character. The argument is that life is now so much easier than it was, that the unintelligent can cope and survive and that under these protective conditions their fertility is greater than that of the intelligent. Presumably in the past, unintelligent people found it hard to get a living or communities of high average intelligence were able to replace peoples of lower intelligence and so intelligence was built up. Nowadays, it is said that selection favours deterioration of intelligence. The reasons for believing this are that large families of brothers and sisters are less intelligent on average than small ones; people in positions one is accustomed to regard as occupied by the intelligent on the whole have smaller families than people in less demanding positions.

I am not convinced by this line of argument. We do not yet know the correlation between the I.Q. of the parent as measured on the parent and the number of his children. All such correlations are inferred from indirect evidence; occupation or the intelligence of the sibship is used to judge the intelligence of the parents. This is unsatisfactory. In addition, calculations of what the selection differential in fact is leave out the bottom end of the intelligence scale, the imbeciles and idiots who are not counted as members of the normal population, but belong to it. To calculate the selection differential one needs two pieces of information (a) the average I.Q. of all those reaching sexual maturity no matter what their physical or mental state, and the average I.Q. of those that do not, and (b) the average I.Q. of all those parents, male and female, giving birth to children in one calendar year, weighted for the number of children born and the average I.Q. of those reaching sexual maturity that do not have children. This figure should be taken for several calendar years. There is no information of this kind in the literature as far as I can see and nothing from which it can be reliably inferred. As a matter of fact, direct measures of intelligence of one population taken at intervals indicate that I.Q. is increasing but there are too many special influences for one to believe any trend to be more than a temporary one. On the one hand, one has all the improvements in nutrition which have hastened growth rate and the onset of sexual maturity. Perhaps a given mental age is attained younger now, and this gives the impression that intelligence measured on succeeding generations of schoolchildren is increasing. On the other hand, the introduction of birth control on a large scale came in during this century and was unquestionably used first by those classes which now have the lowest birth rate. This trend will pass if the technique spreads.

City living, which is becoming commoner, is another factor which needs to be taken into account in interpreting trends. The country is in general more fertile than the town. The suggestion has been that there is some inevitable physiological correlation between fertility and intelligence but this could not be proved without a direct measurement of the intelligence of parents and the sizes of their families. I do not believe there is any such inevitable negative correlation. The size of the families of leading occupations a century or so ago was large. In the early years, Presidents of the United States of America, despite George Washington, who had no children of his own, had families whose average size was 4·4; the average has fallen steadily to 2·5 for the last five Presidents. This is obviously an environmental trend which may be reversed at any time.

Though intelligence cannot be proved to be on the decrease because of a rise in Darwinian fitness of the unintelligent due to the improvements in civilised living conditions, the danger that something of the kind might happen is quite real. A second example might be the occurrence of multiple births. It might be argued that in recent years, mortality amongst twin births relative to singles has improved and that twinning should be on the increase. It is on the increase in Finland, but the trend is disguised by the fall in age of mother. Young mothers are less likely to have twins and as the age of mothers in Finland is falling, the overall rate is not obviously changing; however, for any given age of mother, twinning has been going up at the rate of 0·10 per cent each generation since 1875. Making assumptions about heritability and selection differentials, Mr Brown and I calculate twinning rate could rise from 1·5 to 10·5 per cent in thirty generations.

My conclusion is that the short-term effects of evolution, tidying up the mistakes of mutation, require tackling at once. The attempt would be rewarding and can be undertaken without fear of causing a landslide in the human genotype. The more long-term ambition of affecting the course of human evolution is so long-term that we need not hurry over it. There is time to accumulate much more understanding of what sort of life human beings should lead and what sort of human beings they should be.

Summary

The genetic changes which may have been taking place during the 50,000 generations which have elapsed since the first appearance of Australopethicus are: slow long-term trends with a mean selection differential of $\cdot 004\sigma$ approximately; shorter-term trends due to strong selection pressures such as invasion by a new disease which may be expected to have a selection differential of about $0\cdot 1\sigma$; and trends due to changes in selection pressures which upset pre-existing equilibria between mutation and selection. Gene frequencies which are no longer at equilibrium levels will change by mutation alone at exceedingly slow rates and by selection if there is a selection differential as above. Migration has been ignored.

It is argued that research could usefully be made into the occurrence of genetic abnormalities in the population, which at present amount to some 3·5 per cent of all births, without risk of any deterioration of the human genotype; that medical science cannot have been responsible for any deterioration of the human genotype to date; efforts directed to changing the nature of the human population will take such a long time to bring about that they can safely be left until more is understood about their desirability.

References

1 T. Dobzhansky, *Mankind Evolving*, New Haven and London, 1962, Yale University Press, 162-200.

2 United Nations Scientific Committee on the Effects of Atomic Radiation, Report to General Assembly, 13th Session, Supplement 17 (A/3838), 1958. Annex. H., 172-204.

Comments

J. H. BENNETT Probably the most important change in recent times affecting human populations has been the big decline in death rates and the smaller decline in birth rates. We are well aware of the effect of this change on population number, but what of population quality? As Dr Rendel indicated, the main force in evolutionary change is selection acting via differential mortality and/or fertility. The lowered death rate has probably reduced the *opportunity* for selection by death but there are some indications given by Crow[1] that the lowered birth rate may have been accompanied by an *increased opportunity* for selection via fertility. This takes account not only of the lowered mean but also the changed variability in family size. Almost forty years ago, Fisher[2] attached great evolutionary significance to variation in human fertility. He developed an interesting theory to explain the rise and fall of civilisations of which Toynbee describes about twenty before this one of which we are part. Fisher contrasted barbarian groups, where the more successful are also the more fertile individuals, with civilised societies. In the latter, with an economic system where accumulated wealth is influential in determining the social position of an individual and his descendants and the cost of having children is borne by the parents, Fisher maintained that the social selection of infertility would result. I doubt if we are better placed to judge the likely validity of this theory than forty years ago because there have been many special factors affecting fertility differentials in this period such as the spread of contraception at varying rates to different social groups. The recent rapid changes in New Guinea might, however, provide a good opportunity to examine this. Natural selection via deaths has probably been replaced to some extent by natural selection via births. As Dr Rendel suggested, it seems desirable that man should increase *artificial* selection via births through the detection and counselling of heterozygous carriers of harmful genes and in other ways. Natural selection has, of course, not been eliminated nor perhaps drastically reduced in effectiveness. Penrose[3] suggests that in civilised countries like Britain and the United States almost one-half of the zygotes formed fail to reproduce. Selection via embryonic mortality makes an important contribution and this need not be considered harmful; perhaps there are opportunities for artificial selection here which deserve more study. Selection, if it is to lead to change, must have genetic variation to act upon. With intelligence, there seems to be much genetic variation available so that selection should be very effective with significant consequences for society.

Another important change has been in population structure due to the increased mobility of man. The decline in consanguineous marriages and in reproductive isolation of small communities and of different racial groups must have led to an increase in heterozygotes and hybrid vigour. The mating of like-with-like is much more common than in the past, partly because universities and other specialised institutions such as, for example, those caring for the blind and deaf, bring together young people with similar abilities or disabilities from a wide area. For characteristics having a genetical component, such assortative mating may well lead to an increase in the frequency of extreme types.

Genetic polymorphism is of common occurrence in man. Harris[4] studied ten arbitrarily chosen enzymes and found three striking examples of polymorphisms and he suggests that this proportion is typical for all genes in man. Common variants may well involve large selective differences and so provide excellent material for studying how population frequencies and selective values are affected by changed conditions of life. A striking example is the decline in frequency of the sickling gene following malaria control. There are many fascinating questions in human evolution at the present time. What is the genetic effect of changes in the diet or exposure to new chemicals, especially drugs? What are the genetic consequences for the population of the control or treatment of diseases such as smallpox, tuberculosis, and diabetes? Why is colour blindness much more common in Europeans than in primitive cultures and to what extent does a colour-blind citizen today experience death selection in a big city? What is the genetic component in human behaviour and special abilities and what is the role of selection here?

Man's genetic endowment has been moulded by natural selection acting in the past under quite different conditions from those which now exist. Natural selection remains an important factor in determining the genetic constitution of future generations but man is now acquiring knowledge which will enable him to influence this in a predictable manner. Thus many other questions are prompted by Dr Rendel's paper. For example, are there desirable changes which should be made in the recording of birth, marriage, morbidity, and mortality data? Should we have a registry of certain genetic traits as in some other countries? What should be the place of genetics in medical education in Australia and of human biology in general education? I am sure that Dr Rendel's excellent paper will stimulate much thought.

[1] J. F. Crow, 'Mechanisms and Trends in Human Evolution', *Daedalus* (Summer), 1961, 416-31.
[2] R. A. Fisher, *The Genetical Theory of Natural Selection*, Oxford, 1930, 2nd ed., 1958, Clarendon.
[3] L. S. Penrose, 'Natural Selection in Man' in *Natural Selection in Human Populations*, D. F. Roberts and G. A. Harrison (eds.), 1959, Pergamon Press, 1-10.
[4] H. Harris, 'Enzyme Polymorphisms in Man', *Proc.Roy.Soc.B.* **164**, 1966, 298-310.

D. G. CATCHESIDE Dr Rendel's account of the time scale of genetic change is a valuable summary which would allow us to calculate the genetic changes in man which may have occurred in the past few thousand years if we knew the parameters to use. Participants in this symposium are interested in whether significant genetic changes in man have accompanied his change from palaeolithic culture to our present civilisation and will have noted Dr Rendel's opinion that, with the advent of civilisation, there has been no acceleration in man's physical or intellectual evolution. Whether this is true is a crucial matter, but I doubt whether we have the evidence to judge. Is there any reason to suspect otherwise?

It would be difficult to support the view that no genetic change has occurred, for this would be equivalent to saying that no change has occurred in the selective forces which have operated. Slow changes occurred over the 450 or so millenia (or nearly 25,000 generations) from Java and Peking man to *Homo sapiens*. This is manifested in a steady improvement in skill in the manufacture of tools and in the development of other arts. In general, the genetically better endowed people would have had an improved chance of survival and so there would be a steady change in heredity. There would be a general selection of favourable combinations of genes to change the composition of the gene pool.

From the time of the invention of agriculture, which made civilisation possible and indeed necessary, a relatively short time has elapsed, perhaps 400 generations. The invention of agriculture would have improved the survival of the inventors, permitting a marked relative increase in population from this source and because of the settled character of the population imposed by the agricultural activities there would be a reinforcement keeping the favourable heredity in the local population. In time this would explode outwards and the agricultural, peasant skills would spread widely. We know in historic times how readily technically advanced groups displace backward ones. There would no doubt be interbreeding with other local populations encountered, but with further reinforcement of the agricultural skills. Similar effects could occur with any other selectively advantageous skill having a genetic basis. The invention of agriculture etc. by an isolated group would be equivalent to the eruption of a genetically isolated population into a new ecological niche.

It is difficult to assess whether the time scale, measured in generations, has been sufficient to allow very substantial changes in genotype even with strong selective advantages, though some think so. Manifestly, the populations in different parts of the world show considerable genetic differences. Do these represent differences due to divergence of the separate populations over the 200,000 years during the evolution of *Homo sapiens* or are they mainly the product of chance divergence during the spread of agricultural, civilising man or a mixture of these processes?

Dr Rendel thinks there has not been any great change in man's intellectual capacity over the past few thousand years. But were the Greeks rich in scientists, mathematicians, and dramatists because they were descendants of a few inventive geniuses who developed agriculture, baking, brewing, pottery, and metallurgy? Has cultural evolution occurred through the selection of favourable heritable changes or in spite of an absence of change? We do not have the evidence to decide. We need to measure and understand the genetic diversity of human populations and how this has evolved, whether over 500,000 or 200,000 or 10,000 years.

Discussion

FREEMAN I am in agreement with Professor Catcheside that research on the aetiology of human evolution is crucial for the understanding of modern man and his adaptations. Our ignorance in this field is vast, but there are now abroad, in anthropology, some new realisations. If we consider the spectacular changes from the beginning of the Pleistocene to the present, to which Dr Rendel has drawn our attention, human evolution, it is evident, has proceeded at a comparatively rapid rate, in response to selective pressures which we are, as yet, unable to identify. In my opinion, we are ill-advised to suppose that these pressures were wholly ecological. Savannah-based predation was clearly of prime significance during the early Pleistocene, but this environmental adaptation, it is important to realise, was sustained by concomitant cultural adaptations. And so our forbears entered a novel evolutionary phase in which cultural adaptations, as such, began to exert selective pressures. In other words, there was a selective advantage conferred on those populations capable of producing individuals with the capacities (essentially genetically based) to sustain cultural innovations of an adaptive kind. Thus, cultural adaptations are beginning to be seen as intrinsic to human biological evolution, and in intimate communication with the genetically-based evolution of human populations. If we further consider the human evolutionary sequence, it is conspicuous that there has been exceptionally rapid evolution, in adaptational terms, since the Mesolithic. It is often assumed that this rapid evolution has been wholly cultural, and entirely without any evolutionary change of a genetical kind.[1] I would like to question this assumption. It is now evident that the Mesolithic and the Neolithic extended over a hundred and more generations, and if the genetic character of the human populations in those regions where rapid evolutionary change did occur was sufficiently heterogeneous and so genetically pre-adapted to respond to the strong selective pressure of such major cultural innovations as the invention of agriculture, it is very possible that there may have been significant adaptations, of a partly or even predominantly genetic kind, in the human populations involved. Dr Rendel's schema, which is of a classically genetical kind, does not, it seems to me, pay sufficient regard to the decisively important and uniquely historical selective pressures exerted by cultural innovations on certain human genetic pools from the Mesolithic onwards, and on the consequences of these events for the understanding of the time scale of genetic change in human populations. I would, then, like to ask Dr Rendel to comment on the views I have just expressed. It seems to me that if we are to advance our comprehension of the aetiology of human evolution there must be more collaboration between geneticists, prehistorians, and anthropologists.

[1] Cf. the opinion of Ogburn that 'we have no evidence that the biological element in society has been evolving during the past twenty-five thousand years' (W. F. Ogburn, *Social Change with Respect to Culture and Original Nature*, New York, 1950, Viking Press, 376). Cf. also, the statement by Professor René Dubos (in the course of his address to the Royal Society of Canberra on 11 September 1968) that, in his view, there have been no genetic changes in the human species during the last one hundred thousand years.

RENDEL I do not think that what I have said disagrees with Dr Freeman's attitude in any essentials. I do not agree that my schema is classically genetical; it uses the classical approach towards the change in frequency of single genes; it also uses the approach of the quantitative geneticist and animal breeder in so far as changes in phenotype which are not attributable to single genes are concerned. Nor do I agree that pressures exerted by cultural innovations are uniquely historical. All evolutionary processes are historical, and depend on what has gone before; genetic changes of a kind which make cultural changes possible take time to have their full effect since one is concerned with changes in the organisation of a whole society, in addition to those in the development of a single individual; clearly Watt's work must precede Stephenson's but Stephenson's work can follow Watt's without any change in genotype. This is true in some degree of all evolution. Cartilage precedes cartilage bones and so on.

The main comment of Dr Freeman is one with which I am in full agreement. Cultural factors make up one of the sources of selection pressure which are important in human evolution, and I have not ignored them. In fact (on page 28) I list changes 1, 2, and 3—all of which could be affected by selection pressures exerted by culture. Culture is one component of the environment of the human genotype and I explicitly state this in 2 and implicitly in 3. I agree that the selection pressures of culture require careful study and that human evolution will never be understood until these pressures are understood. I have not tried to analyse them. I have been talking of rates of change. Assuming cultural factors can exert as strong a selective pressure on who will and will not breed as does the plague, one might expect them to bring about a change in the quality through which they act of one standard deviation in 1,000 years. I do not know, but doubt whether they will often act as strongly as the plague. Disease resistance is by and large a one-directional quality, it leads to survival at one extreme, death at the other; responses to cultural stresses are apt to be more complex. For example, in society there are obvious advantages and disadvantages to the individual who is aggressive, but it is possible that the meek will inherit the earth for sociological reasons. The balance in society between the advantages of being meek and being aggressive is likely to be closer and lead to smaller selective advantages one way or the other than the ability to recover from the plague in times when that disease is rampant.

MIMS I want to refer to this important distinction between what has been genetically, as opposed to culturally, acquired in human history. It is a strongly charged subject, emotionally and politically. When we ask whether human populations differ in intellectual capacity as well as in skin and hair we cannot help thinking of Hitler, and nowadays it is difficult to talk dispassionately about the Negro problem in these terms. Surely, rather than ask old-fashioned questions about nature as opposed to nurture, heredity as opposed to environment, we should instead ask about the relative importance of *each* of these factors.

Going back to Sir Macfarlane's opening remarks about smoking, one must agree, but I feel there is something missing. So far there have been few studies on the reasons why people smoke. Are there fundamental human needs which smoking satisfies and which would be left unsatisfied if all smokers stopped smoking? It is easy to forget this side of the problem.

DUBOS I would like to formulate my question around your statement concerning Neanderthal man. You state, as I believe is universally accepted, that Neanderthal man was a highly developed person, made well-shaped tools, had a large brain and yet was rapidly displaced by other kinds of people. I believe that a similar situation occurred during the seventeenth, eighteenth, and nineteenth centuries when European man occupied approximately half of the earth. I would like to ask you to what extent what has happened is not only an evolutionary change, but a displacement of one type of population by other types of populations, and to what extent, if it has happened on several occasions, will it affect your mathematical treatment of the changes that have occurred in man?

RENDEL Mind you, this is not my special subject, but I believe that you are perfectly correct in saying that history shows things happen in waves. I do not think this alters my calculation because the people who are the successive waves have had to evolve to the point they have reached. They were not present 1,000,000 years ago, waiting their opportunity.

We come back to the distinction mentioned by Dr Mims which I would like to elaborate on. What animal breeders and plant breeders are concerned with is the extent to which the variation that they observe in their populations is genetic, and amongst the genetic differences how many are straightforward additive effects and how many have complicated interactions; the variation that you observe at any one moment is partly genetic and partly non-genetic, and you have to distinguish that in order to make calculations and predictions about genetic change. But there is also the other thing; alongside the increase in innate ability of the individual are all the things that he learns which are also passed on from one generation to another—so a whole culture or tribe or civilisation which has got a slight start on another one may be genetically no better, but technically so much better that it can wipe it right out; it is superior because it has had an extra one hundred or one thousand years to learn the techniques and make the developments, and so I think that there is a double distinction to be looked for between one culture and another. First, to what extent has one culture, compared to another, learned and passed on from generation to generation the things it is innately capable of? Second, what was the genetic potential of the one culture compared to the other?

TRACEY In 1,100 years the plague bacillus would have passed through a very considerably larger number of generations than man. Might this not perhaps give a possible, or even a more probable explanation of the variation in mortality over that period?

RENDEL Perhaps Professor Fenner will answer that question.

FENNER I think, as Dr Rendel quoted the situation, at a relatively recent point in time there were differences between the mortality of the presumed selected European population and presumed unselected populations of 35 per cent and 75 per cent. If this is so it suggests that the plague bacillus itself had not undergone major attenuation in this time. But there is other evidence from studies of the resistance of rats in cities of India where plague has not occurred for a substantial period of time. This shows that selection for resistance over the past sixty years is considerable. In challenge experiments one gets mortalities of about 10 per cent in rats in cities of India where plague has occurred regularly and recently, but where there has been no plague in the past eighty or one hundred years the mortality is about 90 per cent. I think that there is strong evidence that the changes have been primarily in the host and not in the plague bacillus, on a global scale.

BURNET (*Chairman*) There is one comment that I should like to make in regard to Dr Mims's contribution—the question of genetic or psychological factors in relation to cigarette smoking. Eysenck has suggested that the people (a) who continue to go on smoking, and (b) who get lung cancer, are predominantly extrovert in temperament, as defined by his set of tests. I do not know how far attempts have been made to see whether genetic factors are responsible for the traits that Eysenck calls extroversion and introversion but I fancy they will turn out to be important. Eysenck is probably right that this temperamental character has a bearing on whether you do or do not continue smoking cigarettes when you have become aware of the effect of the habit on the death rate.

It is a rather characteristic example of what is being stressed throughout the discussion, how difficult it is with a genetically highly diverse species like man, to sort out cultural and environmental factors from those that are genetic. Even in assessing the harmful effects of cigarette smoking we may need to look at the genetic aspect as well as the obvious cultural one.

3 Frank Fenner

The Effects of Changing Social Organisation on the Infectious Diseases of Man

In contrast to other types of disease (genetic, traumatic, degenerative, neoplastic), the infectious diseases are dependent upon contact, either directly, or indirectly through vectors or fomites, between individuals of the same species or, in the zoonoses, individuals of different species. For this reason social organisation, particularly community size, and the degree and frequency of contact between individuals of the same and different communities, has played a major part in determining the nature and prevalence of the infectious diseases of man.

CULTURAL CHANGE AND COMMUNITY SIZE

Several aspects of man's evolutionary history have been discussed in this symposium by Dr J. M. Rendel and Dr Frances Barnes; from the point of view of infectious diseases the most important features of man's cultural development are the size of the individual communities of men, the number and proximity of such communities, and the extent of movement and interchange between them. Table 3:1 illustrates tentatively the time scale of significant cultural events in man's history in relation to community size, and the number of generations that have been available for human genetic changes to occur.

Throughout palaeolithic times (and in modern palaeolithic communities) man lived in small nomadic bands of up to one hundred individuals who rarely made prolonged contact with other such bands. With the discovery and development of agriculture, larger more settled communities developed, but these often moved slowly over the countryside as dictated by agricultural techniques (e.g. slash-and-burn agriculture). The great change in community size dates from the development of irrigated agriculture, some six thousand years ago.

The major changes in the social organisation of man have therefore occurred during the last few thousand years of his history, and especially in the last few hundred years since the introduction of steam power initiated the change from rural to urban man. Examples of almost all the forms of human social organisation that we shall consider in a historical sense are with us now in some part of the world or another; the isolated bands of foodgatherers and hunters characteristic of palaeolithic man, the primitive agriculturists and nomadic pastoralists, the 'primitive' unsewered cities of early industrialisation and modern

Table 3:1

The time scale of cultural changes in man, in relation to the number of generations and the size of human communities

Years before 1968	Generations	Cultural state	Size of human communities
1,000,000	50,000	hunter and foodgatherer	scattered nomadic bands of <100 persons
10,000	500	development of agriculture	relatively settled villages of <300 persons
5,500	220	development of irrigated agriculture	few cities of 100,000; mostly villages of <300 persons
250	10	introduction of steam power	some cities of 500,000; many cities of 100,000; many villages of 1,000 persons
130	5	introduction of sanitary reforms	
0	—	—	some cities of 5,000,000; many cities of 500,000; fewer villages of 1,000
		Tropical Africa	
1,000,000 to 2,000	50,000 to 100	hunter and foodgatherer	scattered nomadic bands of <100 persons
2,000 to 50	80 to 2	slash-and-burn agriculture	villages of <300 persons, slowly shifting; many within mosquito range of each other

sophisticated cities of our affluent Western society. We shall draw on this contemporary material in trying to understand the historical changes as they relate to infectious diseases.

THE INCIDENCE OF LETHAL INFECTIOUS DISEASES

As background information, it may be worth noting the present estimated world-wide incidence and mortality due to infectious diseases. Le Riche[1] has estimated that in 1963 there were some 50 million deaths in a world population of about 3,000 million; one-third of these were due to infectious diseases. The death rate varied enormously; it was 1 in 2 or 3 for smallpox and cholera, 1 in 20 for typhoid, and 1 in 250 for dysentery. Some of the most prevalent infections, the mild viral infections of the upper respiratory tract, carry no mortality. Tuberculosis, in 1963, caused some 3 million deaths, measles half a million. In addition to this background of endemic disease, epidemic and pandemic spread of a few diseases still occurs, notably influenza which caused some 700 million cases and 20 million deaths in 1918-19, and even more cases, but many fewer deaths, in 1957-8.

EXAMPLES OF THE EFFECTS OF SOCIAL CHANGES ON INFECTIOUS DISEASES

In discussing the effects of changes in social organisation on infectious diseases I shall first give brief and speculative accounts of some evolutionary changes in parasite and host in a few specific human diseases: malaria, a mosquito-transmitted protozoal disease; tuberculosis and salmonellosis, two bacterial diseases acquired by inhalation and/or ingestion; and several viral diseases, measles, chickenpox, poliomyelitis, yellow fever, and the common cold. I shall then try to relate types of social organisation with the incidence and severity of infectious diseases.

Protozoal infections

Falciparum malaria. Human malaria can be caused by four species of plasmodia; I shall confine my remarks to falciparum malaria which is by far the most lethal type. *Plasmodium falciparum* is essentially a tropical parasite. Not only does it require a relatively high temperature to develop in mosquitoes, but since there are no persistent exoerythrocytic forms of the parasite to provide the chronicity found in other types of human malaria, transmission to mosquitoes must occur early in human infections.

We are quite ignorant of the origin and evolution of *P. falciparum*. Although it can be transmitted experimentally to some primates, it does not appear to have a primate reservoir and its lethality bespeaks its relatively recent emergence as a major human parasite. In the present context I wish to speak of falciparum malaria mainly to illustrate the complex interactions between this infection and human cultural and genetic changes in tropical Africa.

For the major part of their vertebrate cycle the plasmodia are parasites of the red blood cells and their capacity to multiply in these cells is greatly affected by the kind of haemoglobin they contain. The red blood cells of human foetuses and young infants contain, besides normal 'adult' haemoglobin, the chemically different 'foetal' haemoglobin, which disappears in later life. *P. falciparum* appears to grow less readily in cells containing foetal haemoglobin and young infants appear to be correspondingly somewhat resistant to falciparum malaria. After infancy, in the absence of treatment, falciparum malaria is highly lethal; the tolerance of the infection that is found in highly endemic areas is purchased at a cost of general ill health and a high childhood mortality.

There is good evidence that falciparum malaria has been the selective agent for the spread through certain human populations of several other mutant forms of human haemoglobin, of which the best understood is haemoglobin S. When the gene for haemoglobin S is in the homozygous state there are changes in the red blood cells which lead to severe intravascular haemolysis; the abnormal cells can be recognised in slides by their sickle shape. Heterozygous individuals do not suffer from this haemolytic disease. Parasitaemia and mortality due to falciparum malaria are consistently lower in individuals with haemoglobin S, whether they are heterozygotes or homozygotes,[2] and the distribution of falciparum malaria and the sickle-cell trait show a remarkable overlap, especially in tropical Africa.

Wiesenfeld[3] has produced an interesting hypothesis relating the selection of the sickle-cell trait, by way of falciparum malaria, with the introduction into Africa of what Murdoch[4] has called the Malaysian agricultural complex. This consists of the cultivation, after slash-and-burn destruction of the jungle, of yams and taro as root crops, and bananas and coconuts as tree crops. Before the introduction of this cultural change, the inhabitants of tropical Africa (other than the Pygmies) were for the most part restricted to the sea coast and the large water courses. Introduction of these crops allowed them to penetrate the tropical rain forests and to reach much higher population densities than the Pygmies who were living there as hunters and foodgatherers. The adoption of this agricultural system, which provided a greater and more certain food supply, led to both expansion and concentration of the population. More people were centred in one area and there was less movement, due to the greatly increased needs of husbandry to maintain the yield of the agricultural system. In such a situation, in a malarious area, the incidence of malaria increased, for the probability of transmission increased. Furthermore, slash-and-burn agriculture led to a great expansion of breeding places for mosquitoes of the *Anopheles gambiae* complex, which are highly efficient vectors of falciparum malaria. With more intense malaria there was a strong selection for the originally rare sickle-cell gene (Fig. 3:1), and a consequent increase in the Darwinian fitness of the

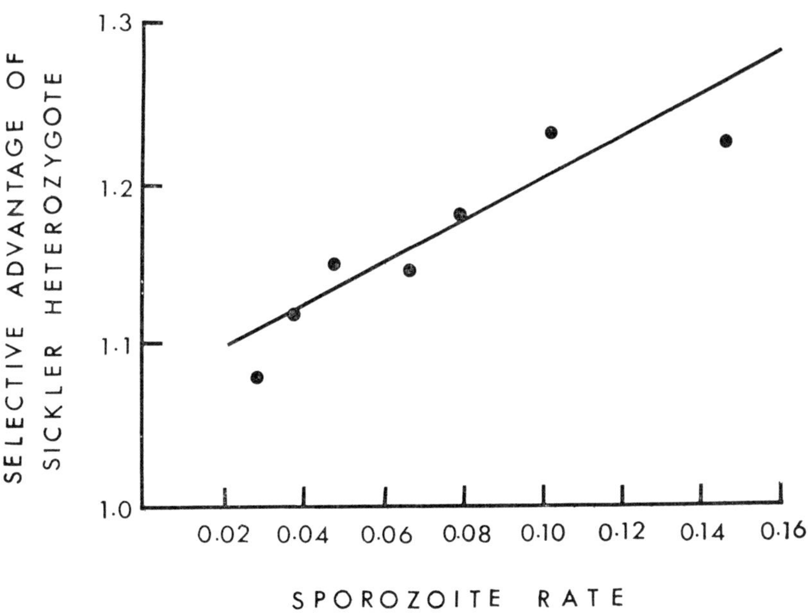

FIG. 3:1 Relationship between the intensity of endemic malaria (the sporozoite rate) and the selective advantage of the 'sickler' heterozygote over the normal population (from 3)

population in these malarious areas. Wiesenfeld has developed mathematical models of the interaction of the sickle-cell trait and malaria which suggest that the sickle-cell trait stabilised in hyperendemic malarious areas in 40-50 generation (1,000 years) at a level very close to that now found in Uganda (Fig. 3:2). In the savannah areas on the fringes of the tropical rain forest, where the humidity was lower and therefore mosquito survival was shorter, equilibrium was reached in

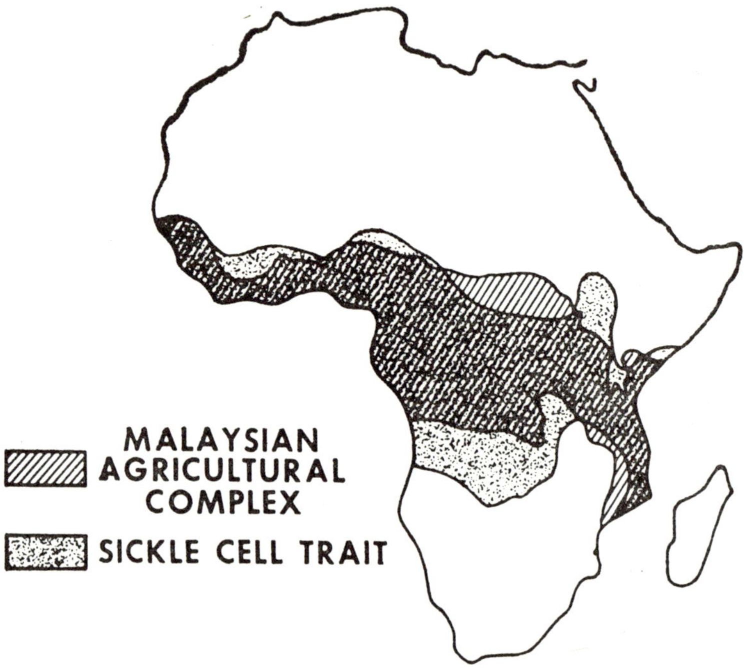

FIG. 3:2 Distribution of the Malaysian agricultural complex and of sickle cell trait frequencies higher than 5 per cent (from 3)

60 generations or 1,500 years at a value close to those reported for north western Nigeria. Haemoglobin S is only one of the factors that is important in resistance to falciparum malaria; other haemoglobin factors, certain red cell enzymes, and probably several other unknown factors also play a role.

City-dwelling eventually eradicates falciparum malaria, for the development of towns usually leads to the eradication of breeding sites. But this sequence is not inevitable; in Kano (northern Nigeria), which is one of the old cities of the world, there are breeding sites in the town itself because the mud used for buildings is obtained locally. In Karachi, quite recently, there was a very large epidemic of 'man-made' malaria in an area which had been malaria-free.

With the disappearance of falciparum malaria, the sickle-cell trait, which causes severe disease in homozygotes, becomes disadvantageous. A new equilibrium will therefore be reached, at a rate which is dependent upon the level of medical care available in the community, for the homozygous can only be kept alive by sophisticated medical attention. The enforced migration, some 5-8 generations ago, of West Africans to North America, where falciparum malaria has always been rare, had led to interesting genetic readjustments which give a preview of what may occur with the great reduction of falciparum malaria that has now been achieved in parts of tropical Africa, by the application of recently developed methods of mosquito control.

Thus we can picture falciparum malaria as a new human disease emerging in the tropics during the palaeolithic period, causing high but not crippling mortalities in the hunters and foodgatherers of the tropical forests, and increasing in intensity as the size of human populations and the opportunities for breeding of vectors increased with advances in agricultural practices. This increased intensity in larger human populations led to rapid selection for traits which conferred increased likelihood of survival and fertility, and a number of mutations like haemoglobin S and 6-glucose dehydrogenase probably arose independently, on more than one occasion, in different parts of the range of *P. falciparum*. Falciparum malaria in Africa provides a good example of the role that disease can play in human evolution, with human biology and culture interacting and differentiating together in stepwise fashion.

Bacterial infections

Tuberculosis. Two types of tubercule bacillus commonly affect man, the human and the bovine type. In general the former is much the more important, and I shall be concerned primarily with it. Tuberculosis due to the bovine type tubercle bacillus is almost entirely confined to people who drink contaminated milk, which restricts its incidence to certain human cultures. Western urban man is a milk-drinker, but sanitary methods (tuberculin-tested herds and the pasteurisation of milk) have eliminated infection of man by bovine type tubercle bacilli in most affluent societies. Comparable considerations apply to other infections which can be acquired from the milk of cows or goats, such as brucellosis, Q fever, and Central European tickborne encephalitis.

Tuberculosis due to the human type of tubercle bacillus has a totally different ecology and a fascinating history, which has been described at length by our American guests at the symposium, Dr and Mrs Dubos.[5] Tuberculosis due to the human type bacillus is for the most part a respiratory disease. Infection is due to inhalation, and bacilli are ejected by coughing, sneezing, spitting, etc. Tuberculosis is often a very chronic disease, and this chronicity offers a means of survival for the tubercle bacillus which may span human generations and thus allow it to persist in small closed communities. Individuals who may not be sick enough to be segregated in hospitals may excrete tubercle bacilli daily for years; in other cases viable bacilli may persist in 'cured' lesions, and may be excreted when these are reactivated by some environmental stress or concurrent infection.

The long persistence of the bacillus in a viable state and its persistent or periodic excretion ensure survival of the tubercle bacillus even in sparse populations. The probable existence of tuberculosis amongst pre-Columban Indians in North America suggests that it was an ancient disease of man, possibly even of his pre-hominid ancestors. However, the dimensions and importance of this 'palaeolithic' tuberculosis may be altered out of recognition by social changes. Those we know most about are comparatively recent, in the miserable crowds of the industrial cities of the Western world in the eighteenth and nineteenth centuries, and in the modern slums of the new industrial cities of Asia and South America. Under these conditions tuberculosis became a major epidemic disease, and amongst those human groups longest exposed to its rigors there was a selection for genetic resistance. Tuberculosis in the Ashkenazi Jews, and in Europeans in general, is usually much less acute than amongst those first exposed to it, like Africans and South Sea islanders; but even this resistance is at best partial, and can be overcome by starvation and physiological misery of the type to which the Jews were subjected in Europe during the last World War.

The amelioration of tuberculosis in the cities of Europe was due in part to the elimination of some of the genetically most susceptible by early death, but in larger part to the sanitary revolution ushered in by Chadwick, which led to such an alteration in the environment that transmission became less frequent and dosage less massive. At about the same time nutritional and other improvements enhanced resistance to the disease. Specifically medical measures took over where the sanitary revolution stopped, so that in affluent Western societies tuberculosis has once again become relatively rare as a crippling and killing disease.

Salmonellosis. The term salmonellosis describes the infection of vertebrates by one of a variety of closely related bacteria, some of which may cause a localised disease of the gastrointestinal tract, whereas others, like the typhoid bacillus, may cause severe generalised infections. All are acquired by eating or drinking and are excreted in the faeces, and sometimes the urine, of man or his domestic animals. With rare but notable exceptions (like 'typhoid Mary') they cause acute diseases in which excretion of bacteria in the faeces is restricted to the few weeks during and after the acute infection. Such agents would not readily survive in scattered nomadic communities. In settled agricultural and town communities with a communal water supply and a limited understanding of hygiene, infection is very common, and infection during childhood leads either to death or to the development of immunity in later life. The apparently healthy adult communities are often heavily infected with salmonellas and other enteric pathogens, including hepatitis viruses and polioviruses, dysentery bacilli, and a variety of worms, as became apparent when large 'clean' populations of European soldiers were billeted near Middle East villages in the two World Wars. In modern cities the development of effective sanitation has ensured that most infected excrement is disposed of rapidly and safely, and in such cities measures are taken which ensure that the communal water supply is not contaminated with bacteria pathogenic for man.

Salmonella food poisoning, as a disease of modern Western man, is a zoonosis, i.e. it originates from some animal source. In contrast to the exclusively human salmonelloses like typhoid fever, which have almost disappeared with the introduction of safe water supplies, this type of salmonellosis is increasing in most Western countries. Four reasons have been suggested for this increase.[6] Firstly, international trade in human foods and animal foodstuffs has increased, and microbes which occur and can survive in such material have thereby become cosmopolitan. Secondly, there is a higher incidence of salmonellosis in farm animals, partly due to the importation of foodstuffs and partly to modern intensive farming, so that more animals which are carriers of salmonellas are slaughtered and there is more contamination of meat. Thirdly, the widespread use of household detergents has interfered with the effective processing of sewage, and effluents from sewage plants now commonly contain salmonellas. Fourthly, the industrial centralisation of food processing and manufacture accompanied by wide distribution of half-preserved or frozen 'prepared' foods has extended the potential range of outbreaks of foodborne diseases, so that cases from one source may be scattered over a very wide area.

Again we can trace the complex interaction of disease and society; at first the rare salmonella infection of the palaeolithic hunter, acquired from some animal that he had killed, then the widespread occurrence of specifically human enteric infections when villages developed with communal water supplies but no understanding of hygiene, and finally the virtual eradication of these diseases in the cities of affluent societies by sanitary measures. In salmonellosis very recent new practices in food preparation and distribution have led to an increase in infections acquired from animal sources, but as these sources are recognised they are brought under control.

Cholera. Cholera is a specifically human disease due to a bacterium which is spread by the faecal-oral route, usually by the contamination of drinking water. It must have been unknown amongst nomadic palaeolithic man, but developed as villages and a village water supply were established. We tend to think of cholera as a 'tropical' disease, but this is due to the current association of poverty and contaminated water supplies with modern communities living in tropical areas. Cholera and other less dramatic enteric infections ravaged the cities of Europe in the early stages of the industrial revolution, during the first half of the nineteenth century, and the greater mortality in the developing cities (expectations of life in 1841 of 36 years for London and 26 for Manchester and Liverpool, compared with 41 for England as a whole) was largely due to cholera and other enteric bacteria, and tuberculosis.

Viral infections

Measles. Measles is a specifically human acute self-limited disease, recovery from which is followed by absolute and lifelong immunity. It is almost unique among the common viral infections of man in that virtually all cases of infection produce obvious symptoms. Virus is excreted for a few days only, at the height of the disease, and neither clinical recurrences nor recurrent viral excretion occur. As we should expect, a disease with these characteristics would be able to survive in small communities only if there were an extra-human reservoir, and no natural host is known for the measles virus except man. Experience with remote and isolated small communities testify both to the disappearance of the disease, if it is not continually reintroduced, and to the lifelong protection afforded by an attack.[7]

The epidemic pattern and the persistence of measles have been further explored by Bartlett and by Black (see [8]). At least thirty cases a year would be needed to maintain the disease if they were evenly spaced and there were no 'misses' in case-to-case infection. Bartlett calculated on theoretical grounds that in fact about 2,500 cases a year are required, in the absence of re-importation. His data from British and American cities suggested that the actual minimum figure needed to prevent 'fade-out', i.e. failure in the transmission cycle, was 4,000-5,000 a year, a number which did not usually occur in cities with populations of less than about 300,000. Island communities provide even better material to study the effects of population size and dispersion on the endemicity of measles. Table 3:2 sets out some data obtained by Black.[9] Measles has disappeared at intervals, and later been reintroduced, in all islands with a population size of less than half a million, and the effect of population size on the frequency of fadeout is clear from the figures. Guam and Bermuda, which form

Table 3:2

Endemicity of measles in islands with populations of 500,000 or less, all of which had at least four exposures to measles during 1949-64 (data from 9)

Island	Population	Annual population input*	% months with measles (1949-64)
Hawaii	550,000	16,700	100
Fiji	346,000	13,400	64
Samoa	118,000	4,440	28
Solomon	110,000	4,060	32
French Polynesia	75,000	2,690	8
Guam	63,000	2,200	80
Tonga	57,000	2,040	12
Bermuda	41,000	1,130	51
Gilbert and Ellice	40,000	1,260	15
Cook	16,000	678	6
Falkland	2,500	43	0

* 1956 births less infant mortality.

exceptions to the pattern, are exceptional also in the much greater inflow of military or civilian transients from mainland USA.

Apart from reintroduction, the pattern of dispersal of the population will obviously have an effect on the endemicity of measles, since the duration of individual epidemics is correlated inversely with population size. Fig. 3:3, constructed for seven island communities of comparable population size, suggests that with maximal crowding an epidemic would be spent in about four months in a population with an input of 1,900-4,500 susceptibles a year.

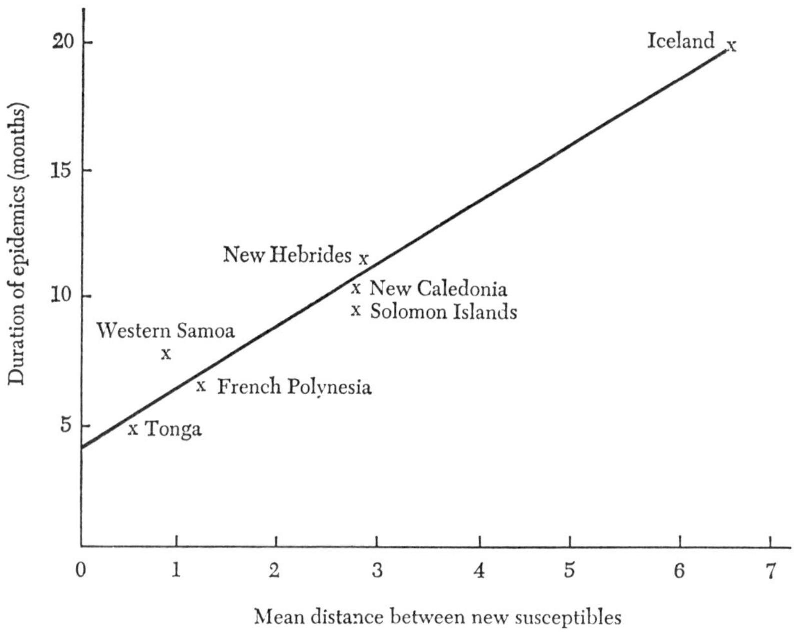

FIG. 3:3 Relation between the average duration of measles epidemics and the degree of dispersion of populations in island communities with about 2,000-4,000 new susceptible children per year. The abscissa plot

$$\frac{1}{\sqrt{\dfrac{\text{input}}{\text{kilometres}}}}$$

represents the mean distance between new susceptible persons (from 9).

I have said that measles is a specifically human disease, yet it obviously could not survive in the sparse communities which existed prior to the introduction of irrigation as a method of increasing agricultural yield; on our time scale, earlier than some 6,000 years or 240 generations ago. Where could it have come from? Two possible sources suggest themselves, canine distemper and rinderpest. These diseases are caused by viruses closely related to the virus of measles. Canine distemper and rinderpest viruses themselves, like measles virus, show fairly high host specificity and long-term immunity, but the much more rapid turnover of canines and bovines ensures their survival. Speculatively, we could ascribe measles to a mutation in distemper virus such that it became a human pathogen, occurring at some time after the establishment of the first cities, some 6,000 years ago. If such a mutation had occurred earlier it would not have survived in man as the sole host.

Previously unexposed human populations have suffered high mortalities when measles first appeared amongst them. This was in part due to the high incidence in adults, who suffer more severely than 5-12 year old children, and partly no doubt to the social disorganisation that accompanied the explosive epidemics. The lack of genetic resistance may also have played a part, for it is a fair presumption that the continued exposure of urban man over many generations has eradicated the genetically highly susceptible individuals from such populations.

Chickenpox. Chickenpox is a specifically human viral infection which, like measles, gives rise to lifelong immunity to reinfection. In striking contrast to measles, however, the critical community size needed to ensure the survival of the disease is very small, less than 1,000 instead of 300,000 to 500,000.[10] This is explained by the fact that varicella virus, after being latent in the human body for many years, may be reactivated. It then causes not chickenpox, but a different clinical syndrome called shingles or herpes zoster. The vesicles of zoster contain chickenpox virus, which is infective for susceptible children, who get chickenpox, not zoster. Although zoster is not as infectious as chickenpox (attack rates of 15 per cent, compared with 70 per cent for chickenpox), it is only necessary to produce one case of chickenpox from a case of zoster to produce a minor epidemic of chickenpox in a susceptible community.

Chickenpox could therefore be a disease as ancient as man himself, and one might expect that it would have been widespread throughout the world before the great explorations of European man, although by accident it might well have died out in some small communities.

Smallpox. Smallpox is a specifically human disease giving lifelong immunity which, like measles and unlike chickenpox, is not subject to recurrences. Like measles, therefore, we can postulate that it must have originated from some animal host after the development of the first cities. Molecular hybridisation techniques, already applied to cowpox, mousepox, and vaccinia virus,[11] could provide a tentative answer to the problem of which animal poxvirus was the ancestral virus of human smallpox.

The case-mortality of smallpox was, and still is, so high that it is likely that selection for genetic resistance has occurred amongst the inhabitants of Europe, India, and China, who have been exposed to the disease for centuries. Certainly the mortalities suffered by previously unexposed populations of Central and North American Indians when they were first exposed to the disease were tremendously high—the intentional spread of smallpox by the English inhabitants of North America was probably the earliest example of deliberate biological warfare. There have been suggestions that smallpox resistance may be correlated with certain blood group antigens but the published evidence is not completely convincing.

Yellow fever. Yellow fever,[12] which now occurs only in tropical Africa and tropical America, was once thought to be a peculiarly human disease, transmitted from one person to another by the urban mosquito

Aedes aegypti. It is now clear that, like most of the arbovirus infections of man, yellow fever is zoonosis. The natural host animals are African primates, in which infection is asymptomatic. In Uganda, in Central Africa, jungle yellow fever (in man) involves two vector mosquitoes. *A. africanus* is a forest canopy mosquito which maintains an enzootic cycle in wild primates. The virus is conveyed to man by a second cycle in which monkeys raid plantations and infect local *A. simpsoni* mosquitoes which then infect man and can maintain a man-mosquito-man cycle.

In Central America jungle yellow fever in man is mainly a disease of the adult male whose work brings him into close contact with the forest, but it may occur among women and children when families are involved in crop cultivation in forest clearings, to which infected monkeys may be attracted as marauders.[13] The principal vectors are mosquitoes of the genus *Haemagogus* which usually inhabit the forest canopy although they may bite man outside and even inside houses.

Jungle yellow fever is a prototype of the sort of viral infection to which palaeolithic man was exposed, and there are many other arboviruses extant in the tropics and subtropical areas which likewise may occasionally infect man, usually producing a dead-end infection. Classical urban yellow fever was clearly a disease of cities and hence of relatively recent origin. Since immunity after recovery from yellow fever is lifelong and there are no recurrent episodes, urban yellow fever was subject to the same limitations for its survival as measles, although occasional long-lived infected mosquitoes might 'tide over' longer intervals than could be tolerated in the case-to-case infection necessary for the survival of measles, and infected mosquitoes may be more mobile than humans infected with measles. However, in the twentieth century this dependence upon mosquito vectors became a weakness, for *A. aegypti*, the vector of urban yellow fever, can be readily eradicated from cities by anti-mosquito measures.

The common occurrence of high mortality in South American jungle primates and severe disease in South American Indians, and low mortality and mild disease in African primates and African man, point to the occurrence of long selection for genetic resistance in Africa. Yellow fever is an African disease which was transported to the Americas quite recently, both the virus and the mosquito *A. aegypti* being taken to America on ships carrying slaves: another example of the interaction of cultural activities and infectious diseases.

Poliomyelitis. Many clinical syndromes, including some generalised infections like poliomyelitis, can be caused by several antigenically different viruses of the same species, three in the case of polioviruses. The disease, poliomyelitis, is a rare complication of poliovirus infection, which is normally a symptomless infection of the superficial cells of the gut. The incidence of the neurological disease depends upon the serotype and the age of the affected individuals, but is usually less than 1 per cent. Like measles, one infection, whether clinical or inapparent, gives lifelong immunity to that serological type of poliovirus. In the absence of re-infection of the gut recurrent excretion does not occur, and there are no animal reservoirs. For these reasons polioviruses could not have survived in palaeolithic man, and there is ample evidence that, like measles, the polioviruses disappear completely from isolated and remote communities like the Eskimos.[14]

In modern human communities determination of the age at which 50 per cent of individuals present serological evidence of past infection with polioviruses provides a useful index of the ease with which faecal-oral viral infections may spread. Thus 50 per cent of the population of Cairo had actively acquired antibodies to poliovirus type 2 by the age of 14 months; in urban areas of USA this point was not reached until the age of 10-15 years.[15]

Neurological complications of poliovirus infection (i.e. recognised paralytic poliomyelitis) are very rare in young infants, but are much more common in adolescents and adults, if such individuals have not been protected by vaccination or prior infection. Epidemics of paralytic poliomyelitis are essentially a phenomenon of the modern world, and have occurred principally in those countries in which a high standard of hygiene has greatly reduced faecal-oral infection.

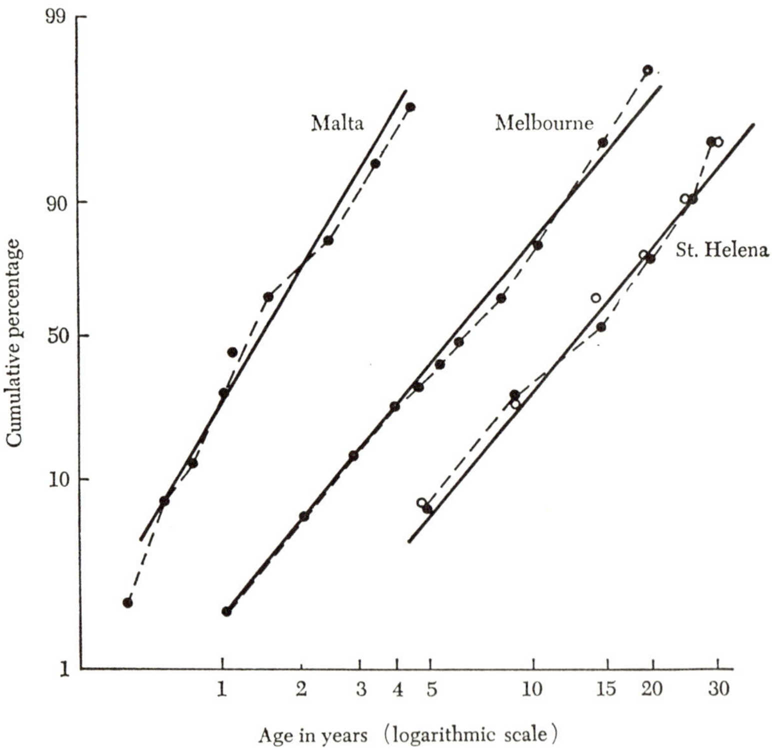

FIG. 3:4 Cumulative cases of poliomyelitis by age in three epidemics. Ordinates show percentage of total cases whose age is less than that shown by the logarithmic scale of the abscissa. Malta, 1942: infantile-type epidemic; Melbourne, 1937: normal modern type; St Helena, 1947: virgin-soil type. Total reported cases, open circles; paralytic type, filled circles (from 15).

Fig. 3:4 illustrates three types of epidemic of paralytic poliomyelitis.[16] The age distribution depends upon the antibody status of the population, the intensity of the epidemic on the intrinsic virulence of the strain of the causative virus and the physiological (age) susceptibility of the non-immune population involved. The Malta epidemic (Fig. 3:4) represents the impact of a virulent virus in a community with low standards of hygiene; in such communities epidemics are rare though infection is common. The Melbourne epidemic typifies the kind of epidemic which was characteristically found in western Europe, USA, and Australia before the advent of vaccination. St Helena represents a virgin-soil epidemic, i.e. poliomyelitis in an isolated community from which the virus had been absent for many years. The effect of age on the incidence of neurological disease in non-immune individuals is clearly seen in the St Helena epidemic and has also been noted in epidemics in Eskimos.

Vaccination against poliomyelitis with two types of vaccine, Salk (inactivated; given by injection) and Sabin (live-attenuated, given orally) have virtually eradicated paralytic poliomyelitis from many parts of the world. Rather surprisingly, in view of the failure of killed virus vaccines to affect faecal excretion of the virus to any marked extent, the circulation of polioviruses in communities well-vaccinated with Salk-type vaccine has also been greatly reduced. A possible explanation is that in communities with a high standard of sanitation (which are those in which Salk-type vaccination has been extensively used) transmission of polioviruses depends mainly on the pharyngeal-oropharyngeal rather than the faecal-oral route. Antibody produced by Salk-type vaccination reduces pharyngeal infection by polioviruses and reduces viral multiplication in that site, thus reducing the transmission of infection. In primitive urban and village communities, in which transmission is by the faecal-oral route, killed virus vaccines have virtually no effect on the circulation of polioviruses, although they may greatly diminish paralysis by interfering with blood-borne spread of the virus through the body. Live-virus vaccines, however, may interfere with the circulation of polioviruses and other enteroviruses at the time of vaccination; and for a prolonged period after vaccination they prevent the implantation of polioviruses in the gut, probably because of local production of IgA antibodies. Mass vaccination with live-virus vaccines thus offers the possibility of interfering with the circulation of polioviruses to the point of at least local eradication, even in communities in which sanitation is poor.[17]

A situation with poliomyelitis is approaching which is not unlike that now prevailing with smallpox, in which rapid travel and mass movements of personnel in war raise possibilities of the re-introduction of virulent viruses into countries from which they have disappeared. With the increasing difficulty of maintaining widespread vaccination campaigns directed towards a locally non-existent disease, a situation may well arise in which evidence of recent vaccination with a live-virus vaccine against poliomyelitis will be a requirement for permission to enter the affluent countries, comparable with the certificates for smallpox or yellow fever vaccination.

Common colds. Superficial infections of the mucous membrane of the upper respiratory tract which cause the so-called 'common cold' may be caused by a very large number of serotypes of rhinoviruses, as well as by respiratory viruses belonging to other taxonomic groups. With very few exceptions the different serotypes of the rhinoviruses are quite distinct antigenically; there is no serological cross-reactivity nor any cross-protection between them. The seemingly endless succession of common colds suffered, usually during the winter months, by most members of modern urban communities reflects a series of minor epidemics due to one or other of the eighty serotypes of rhinoviruses, or to other respiratory viruses.

Colds do not occur in small isolated communities. Perhaps the best example comes from observations made on the inhabitants of Spitzbergen, in the days before air transport existed. As can be seen from Fig. 3:5, there were no colds there throughout the bitter Arctic winter, but an epidemic began as soon as the first boat arrived.[18] It is clear that the same situation holds for any isolated community; most of

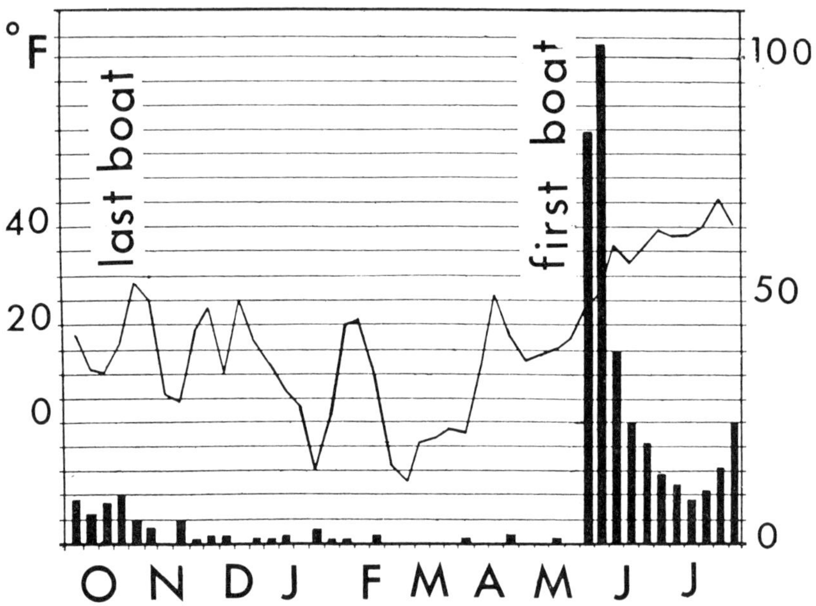

Fig. 3:5 Acute respiratory infections, Longyear City, Spitzbergen, by weeks, 1930-1, with a curve showing the mean weekly minimum temperatures (°F) and the times of arrival of the last boat before winter set in and the first boat after the spring thaw (from 17)

the specifically human respiratory viruses must have evolved since the development of reasonably large aggregations of human beings and regular intercommunication between them.

NEW INFECTIOUS DISEASES IN MAN

The whole pattern of public health and quarantine administration, in relation to viruses, is based upon the assumption that every virus is derived from a pre-existing virus although it may have undergone mutation during the process, i.e. 'new' viruses are not now evolving or developing from non-virus, whether this be cell component or micro-organism. The same assumption applies with even more force to micro-organisms which cause disease in man.

'New' bacterial diseases appear to be relatively uncommon, although human infections with *Pseudomonas pseudomallei*[19] and *Mycobacterium ulcerans*[20] were first recognised as recently as 1912 and 1948 respectively. New viral diseases, on the other hand, have been recognised relatively commonly during this century; most but not all of these have been due to human intervention in some natural situation, or to changes in the social habits of man.

CULTURAL CHANGE AND THE PATTERN OF INFECTIOUS DISEASES IN MAN

In earlier sections of this paper I have described the changing patterns of several infectious diseases of man. Here I shall try to bring out the features of several different infectious diseases which are related to cultural changes in man. Partly because of my better knowledge of them, and partly because of their greater importance for modern urban man, I shall concentrate attention on viral diseases.

Apart from arthropod-transmitted infections, which have special characteristics, the pattern of viral infection sustained by different animals is greatly dependent upon their social contacts with their fellows, and by the life span of individual animals of the species. Man is distinguished from all other animals by the speed with which major changes have occurred in the form of his social organisation; within a few thousand years he has changed from isolated bands of about a hundred hunters and foodgatherers to the vast conurbations of modern man. By comparison, the social organisation of all other animals has been static, except when man himself has deliberately altered the social organisation of domestic animals.

INFECTIOUS DISEASES OF PALAEOLITHIC MAN

Dr Barnes has outlined the sort of social organisation that was characteristic of paleolithic man, modern representatives of whom could be found in a few primitive groups of hunters and foodgatherers until quite recently. From the point of view of infectious diseases the important features of this cultural stage were the small size of the band and its infrequent close contact with members of other closed bands, and the fact that even in palaeolithic times man was a long-lived animal, once the risks of birth and infancy had been successfully passed. Because of their close personal contact and communal habits the microbial flora characteristic of man under these circumstances was probably shared by all members of the group. Many agents, especially viruses, which we regard now as being universal and 'characteristically human', could not have existed. Earlier I discussed the basic community size needed to maintain measles virus; the same situation holds for all the generalised human viral infections for which there is no other animal host and in which latency and recurrent infection do not occur. Immunity due to prior infection is not as durable in many of the diseases of mucous surfaces (intestinal and respiratory tracts) as it is in the generalised infections, but the enteric viruses and respiratory viruses could not survive in man as the sole host in societies of a few hundred individuals. We have ample evidence of the disappearance of respiratory viruses and diseases associated with them, like influenza and the common cold, in isolated communities of modern man; this situation must have been universal in primitive man.

Indeed the only 'specifically human' viral diseases that we could expect to survive in primitive man are those marked by latency and recurrent disease. Herpes simplex virus and chickenpox virus, for example, could survive even in isolated family units, because of this characteristic. Most viral diseases were therefore not 'human' diseases; they were caused by viruses of some other animal which 'accidentally' infected man. The principal diseases of this type are those carried by arthropod vectors, the arbovirus infections, and their incidence in primitive man would obviously differ greatly in different parts of the world, due to climatic differences.

Much the same principles apply to bacterial and protozoal diseases, although especially with bacteria some potential pathogens can survive indefinitely as saprophytes on the surface of the skin or on mucous membranes and invade the body as pathogens when some breach of the surface is produced, by minor or severe trauma. Thus staphylococcal and streptococcal infection of wounds probably occurred, from nasal carriers of these bacteria. Tuberculosis, leprosy, and treponematosis, with their characteristic features of chronicity and recurrent excretion, could survive; and of course primitive man was subject to many bacterial and rickettsial infections acquired from other animals or arthropods, such as salmonellosis, leptospirosis, and flea- and mite-borne typhus.

Infectious diseases in societies based on agriculture

The development of agriculture led to a great change in human habits. No longer were men nomads, and no longer was the population size of a group limited by the availability of natural food. Villages developed, and when irrigation made large-scale agriculture possible men started to live in towns and cities. The pattern of viral infections was greatly influenced by the development of these large societies, for now community size had exceeded the minimum level needed to maintain diseases like smallpox, measles, and rubella, and a large close-knit society permitted the ready spread of faecal-oral and respiratory viruses. The neolithic origins of the 'prototype' rhinoviruses, myxoviruses, and paramyxoviruses remain speculative; almost certainly they were acquired from some animal source. Many animals live in large enough groups to sustain such viruses because their turnover rate (i.e. the accession of new susceptibles) is so much more rapid than in man. Smallpox virus must certainly have been derived from a related virus in an animal host, and we have already suggested that measles virus was probably derived from a primitive distemper virus.

Once the 'prototype' rhinoviruses and enteric viruses had been successfully established in human communities they were subjected to natural selection for survival, which operated, as always, at the level of transmission. Other things being equal, antigenically novel viruses had a better opportunity of becoming established and multiplying to a sufficiently high concentration to be transmitted. This may have been accomplished, initially, by immunological drift, but with the picornaviruses the present picture is that a large number of different serotypes evolved and appear to have persisted. Sampling has not yet been extensive enough, or on a sufficiently large scale, to know whether some serotypes of these viruses actually become extinct and are replaced by new serotypes.

Infection with enteric bacteria was probably widespread in the neolithic villages and towns, for the communal water supply provided, and still provides in some parts of the world, an excellent means for the wide dispersal of bacteria. Tolerated infection of the adults was, and is, purchased at the cost of heavy infantile mortality.

Urbanisation and infectious diseases in the cities

As Davis[21] has pointed out, urbanisation is a recent phenomenon, dating back little more than a hundred years and caused primarily by the recruitment of country folk to the cities. It was initially accompanied by intense squalor and poverty, and a greatly increased incidence of the associated infectious diseases. Some of these, like the water-borne bacterial diseases, were controlled by sanitary measures, introduced in the middle of the nineteenth century; others, like tuberculosis, took a terrible toll especially of the adolescent migrants from rural areas. Tuberculosis is still a major epidemic disease of some of the growing cities in 'poor' nations; in India and South America, for example. In the early days of urbanisation the hospitals themselves were hotbeds of infection with a variety of agents; notably the enteric organisms, staphylococci, and the streptococci that cause septicaemia and pneumonia. Even today bacterial cross-infection in hospitals calls for rigorous controls and constant vigilance.

Ultimately a combination of wealth, with attendant improvement in dwellings and a decrease in crowding, sanitation in the form of a safe water supply and adequate disposal of excreta, and, finally, the development of effective anti-bacterial drugs, have diminished the importance of infectious diseases as causes of death and severe morbidity in the great conurbations that characterise the modern world. There has been no corresponding decrease in respiratory viral infections, which now constitute the major unsolved problem among the infectious diseases. Neither vaccines nor 'air sanitation' offer any prospect of early alleviation of their effects.

Human colonisation and infectious diseases

Until the development of ocean-going ships in Europe, in about the fifteenth century, both urban and rural mankind was largely confined to the continents in which his progenitors had lived as nomads and then as primitive agriculturists. The explosive period of the European colonisation of all the other continents had profound effects on the disease pattern in the European migrants and in the indigenes. European man took his endemic diseases to new virgin populations, and there were explosive outbreaks of measles and smallpox, for example, in the native inhabitants of America and Australia. Tuberculosis took a terrible toll, though in less dramatic fashion. European man likewise intruded on situations in which the indigenous inhabitants had acquired, by natural selection, considerable genetic resistance to diseases lethal to the European, e.g. falciparum malaria and yellow fever in Africa.

Alteration of the environment of man may lead to other major alterations in the incidence of viral diseases. A good example of this is furnished by the history of encephalovirus infections in southern California.[22] The floor of the San Joaquin Valley was largely desert until man transformed it by irrigation into a rich agricultural area. Under natural conditions the encephaloviruses were either absent or persisted only in limited areas, but the irrigation channels, and the vegetation they produced, provided an almost ideal environment for mosquitoes, birds, and the encephaloviruses St Louis and Western equine encephalitis virus. Both these viruses cause disease in man and in one of his domestic animals, the horse.

Air travel and viral disease

As long as travel from the endemic and enzootic centres of disease to disease-free areas was relatively slow, quarantine restrictions operated reasonably effectively in keeping out such diseases as smallpox from the USA and the countries of western Europe. Air travel poses much greater problems, since movement from any part of the world to another takes less than 48 hours, and the volume of air transport is increasing rapidly. In 1954 pandemic influenza spread around the world mainly at the speed of ship and train travel;[23] if another such pandemic occurs one can expect air travel to play a much more important role in speedily seeding the disease in new environments.

Air travel is converting the world of man into one ecological unit. The continuing incidence of serious infectious diseases in the poor and over-crowded cities of the world and the direct and constant communication between these cities and those of the affluent countries poses a constant threat to the latter; to some extent this has been combated by vaccination and quarantine but in the long run the only solution to this problem, as to so many others in the world today, lies in the eradication of the sources of infection by the elimination of poverty.

Summary

From an ecological point of view human infectious diseases can be divided into two groups: those specific for man, and the zoonoses, which man acquires from some other vertebrate host. The incidence of infection with both these types of disease depends upon the relation of man with his environment (including other human beings) and upon what infections occur in different environments. Palaeolithic man, living a nomadic life in small scattered bands, suffered from few of the infections we now regard as specifically human, but was exposed to sporadic infection with agents that normally infected other animals; frequently in the wet tropics, rarely in the deserts. The only specifically human infections which could survive in palaeolithic man were those characterised by chronicity or by recurrences; acute infections followed by recovery and immunity could not have survived.

The development of agriculture, especially where aided by irrigation, led to a rapid increase in the size of the total human population and the size of communities, which settled into villages and towns. Under these new conditions respiratory and enteric microbes were transmitted effectively and the appearance of most of the specifically human infectious diseases dates from this time. Some of the common viral infections, like smallpox and measles, must have arisen during this period by mutation and adaptation of a related virus, probably from a domesticated animal.

The industrial revolution led to urbanisation, i.e. the constant growth in the size of cities by migration from rural areas. Initially this led (and still leads) to overcrowding, poverty, and misery; and even now it is accompanied by a failure of services like water supply and sewerage to keep up with the demands. Urbanisation was therefore accompanied by a great increase in the incidence and severity of the specific infectious diseases of man. With the sanitary revolution, increasing affluence, and the introduction of preventive medicine practices based upon the germ theory of infectious disease, the chain of infection of the enteric microbes has been interrupted. In the affluent societies, tuberculosis has been to a large extent brought under control, and the development of effective anti-bacterial chemotherapy has largely eliminated pneumonia as a cause of death. Both tuberculosis and pneumonia remain scourges in the rural and urban slums of the underdeveloped countries. Viral respiratory infections which cause minor morbidity but carry no appreciable mortality have become more common than they ever were, and there is no immediate prospect of their prevention.

Air travel is converting the world of man into one ecological unit; for long-term protection against the severe infectious diseases these must be controlled everywhere. Minor respiratory viral infections are not susceptible to control by vaccination; their effects could be reduced by effective antiviral therapy when satisfactory chemotherapeutic compounds are discovered.

References

1 W. H. le Riche, 'World Incidence and Prevalence of Communicable Diseases', in Ciba Foundation Symposium *Health of Mankind*, London, 1967, Churchill, 1-50.

2 A. C. Allison, 'Genetic Factors in Resistance to Malaria', *Ann.N.Y.Acad.Sci.*, **91**, 1961, 710-29.

3 S. L. Wiesenfeld, 'Sickle-cell Trait in Human Biological and Cultural Evolution', *Science*, **157**, 1967, 1134-40.

4 G. Murdoch, *Africa: Its Peoples and Their Culture History*, New York, 1959, McGraw-Hill.

5 R. Dubos and J. Dubos, *The White Plague*, London, 1953, Gollancz.

6 W. I. B. Beveridge, 'Animal Health and its Implications for Man', in Ciba Foundation Symposium *Health of Mankind*, London, 1967, Churchill, 63-78.

7 P. L. Panum, 'Beobachtungen uber das Maserncontagium', *Virchows Arch.*, **1**, 1847, 492-512.

8 F. Fenner, *The Biology of Animal Viruses*, Vol. II, *The Pathogenesis and Ecology of Viral Infections*, New York, 1968, Academic Press, 736-8.

9 F. L. Black, 'Measles Endemicity in Insular Populations: Critical Community Size and its Evolutionary Implications', *J.Theoret.Biol.*, **11**, 1966, 207-11.

10 R. E. Hope-Simpson, 'The Nature of Herpes Zoster: A Long-term Study and a New Hypothesis', *Proc.Roy.Soc.Med.*, **58**, 1965, 9-25.

11 A. J. D. Bellett and F. Fenner, 'Studies of Base-sequence Homology among some Cytoplasmic Deoxyriboviruses of Vertebrate and Invertebrate Animals', *J.Virol.*, **2**, 1968, 1374-80.

12 R. M. Taylor, 'Epidemiology', in *Yellow Fever*, G. K. Strode (ed.), New York, 1951, McGraw-Hill, 427-538.

13 P. R. Causey and O. Maroja, 'Isolation of Yellow Fever Virus from Man and Mosquitoes in the Amazon Region of Brazil', *Amer.J.trop.Med.*, **8**, 1959, 368-71.

14 J. R. Paul, J. T. Riordan, and J. L. Melnick, 'Antibodies to Three Different Antigenic Types of Poliomyelitis Virus in Sera from North Alaskan Eskimos', *Amer.J.Hyg.*, **54**, 1951, 275-85.

15 J. R. Paul, J. L. Melnick, and J. T. Riordan, 'Comparative Neutralising Antibody Patterns to Lancing (Type 2) Poliomyelitis Virus in Different Populations', *Amer.J.Hyg.*, **56**, 1952, 232-51.

16 F. M. Burnet, *Principles of Animal Virology*, 2nd ed., New York, 1960, Academic Press, 378.

17 A. B. Sabin, 'Oral Poliovirus Vaccine, Recent Results and Recommendations for Optimum Use', *Roy.Soc.Health J.*, **82**, 1962, 51-8.

18 J. H. Paul and H. L. Freese, 'An Epidemiological and Bacteriological Study of the "Common Cold" in an Isolated Arctic Community (Spitzbergen)', *Amer.J.Hyg.*, **17**, 1933, 517-35.

19 A. Whitmore, 'An Account of a Glanders-like Disease Occurring in Rangoon', *J.Hyg.Camb.*, **13**, 1913, 1-34.

20 P. MacCallum *et al.*, 'A New Mycobacterial Infection in Man', *J.Path.Bact.*, **60**, 1948, 93-122.

21 K. Davis, 'The Urbanisation of the Human Population', *Scient.Am.*, **213** (3), 1965, 41-53.

22 W. C. Reeves and W. McD. Hammon, 'Epidemiology of the Arthropod-borne Viral Encephalitides in Kern County, California', Univ. Calif. (Berkeley) *Publ. Public Health*, **4**, 1962.

23 C. H. Stuart-Harris, *Influenza and Other Virus Infections of the Respiratory Tract*, 2nd ed., London, 1965, Arnold, 101-4.

Comments

SYDNEY D. RUBBO Professor Fenner has traced for us the broad interaction between man and microbes in the changing environments which have characterised man's progress from nomadic hunting groups to the technically conditioned communities of today. In spite of the enormous advances in the understanding, control, and treatment of infectious disease, modern man is still plagued by the challenge of competitive survival with the organisms which he or his domestic animals may carry. Of the 150 communicable diseases listed by the World Health Organization, 35 are due to bacteria, 64 are viral in nature and the remainder are variously caused by protozoa, fungi, and helminths. From this list sixteen disease states might be classed as commonly occurring and of them eleven are caused by bacteria, four are viral and one protozoal in origin. From the global point of view diseases of bacterial and protozoal origin exert their most disabling effect among peoples in the developing countries, whereas in the modern urban environment of Western countries viral infections are at least three times more prevalent than those due to bacteria. However, setting aside geography or social progress, the agents of highest mortality are the bacteria causing pneumonia, gastrointestinal infections, and tuberculosis, which together account for 66 per cent of all deaths due to infection, or 11 million deaths per annum throughout the world. Thus it is apparent that man and microbe have not yet come to a mutually beneficial symbiosis. In the developing countries the infectious load is created largely by bacteria and protozoa, in the advanced societies the problem is one of morbidity from trivial virus infections, mainly of the upper respiratory tract.

In spite of these generalised evolutionary patterns of infections the medical practitioner in modern society is a prolific prescriber of antibiotics. Edmondson[1] concluded from a survey in Australia that in 1966 more than 70,000 prescriptions per 100,000 of the population were issued, a figure which fell neatly on the sharply rising graph for figures from previous years. In England and Wales for the same year some 60,000 prescriptions per 100,000 of the population were dispensed. Fig. 3:6 shows the extent of this prescribing for both countries for three commonly used antibiotics.

One of the complications which flows from this fantastic distribution of antibiotics is the emergence of drug-resistant organisms. The appearance of these new strains of bacteria in the affluent societies illustrates an unusual effect of the changing social organisation on the infectious diseases in man.

It is generally known that when penicillin became freely available in 1946 and was widely used thereafter, a profound selective pressure was imposed on all species of micro-organisms normally carried by man and animals. After about ten years of penicillin therapy there was a marked evolutionary shift towards penicillin resistance in pathogenic staphylococci. These penicillin-resistant strains produced an enzyme capable of destroying the drug, namely penicillinase. For many years the mechanism of selection of these penicillinase producing staphylococci was imperfectly understood, since they could not be produced *in vitro* by the usual techniques of selection. Recently, Novick and Richmond[2] have shown that penicillinase production in staphylococci is determined by the presence of an extrachromosomal element in the cytoplasm, namely a plasmid. A point of great clinical importance which arose from this observation is that penicillin resistance in staphylococci does not develop during treatment by the selection of mutants derived from the original infecting population but by acquisition of the resistant strain from another patient or carrier. Thus, the

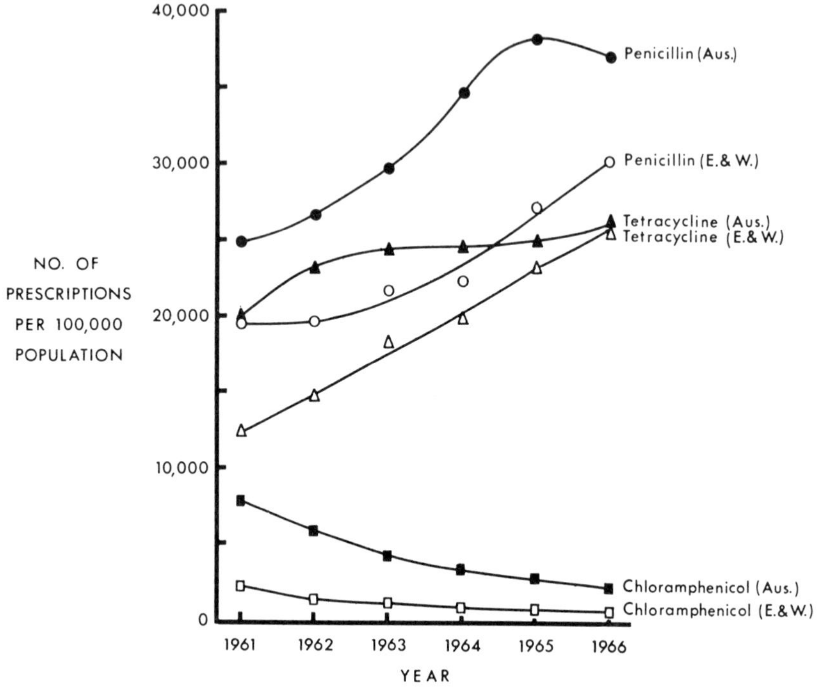

FIG. 3:6 Prescribing of antibiotics in Australia and in England and Wales (from 1)

widespread incidence of penicillin-resistant staphylococci in hospitals is largely the result of cross-infection in a micro-environmental situation which favoured their persistence and propagation.

More recently, a new phenomenon of infectious drug resistance has emerged among the Gram-negative bacilli which are commensals or pathogens of the alimentary tract. Infectious drug resistance first arose in Japan as an epidemiological problem as illustrated in Fig. 3:7. The most frequent combination of drugs to which the organisms acquired resistance were sulphonamides, tetracyclines, chloramphenicol, and streptomycin, all dissimilar in their modes of action and presumably distinct with respect to the genes determining resistance to them.

The curious phenomenon is now fairly well understood and results from 'infection' of a sensitive cell after physical contact with a donor cell carrying an R factor. This R factor is a plasmid carrying a varying number of drug resistance markers. R factors may have a high mobility from one cell to another and may pass from a non-pathogenic species, e.g. *Escherichia coli,* to a pathogen such as *Salmonella typhimurium.* The importance of these R factors, presumably selected by excessive use of antibiotics, must not be under-estimated. From the prevalence of viral infections in the antibiotic era and the emergence of drug-resistant bacteria, we can extrapolate and suggest that our

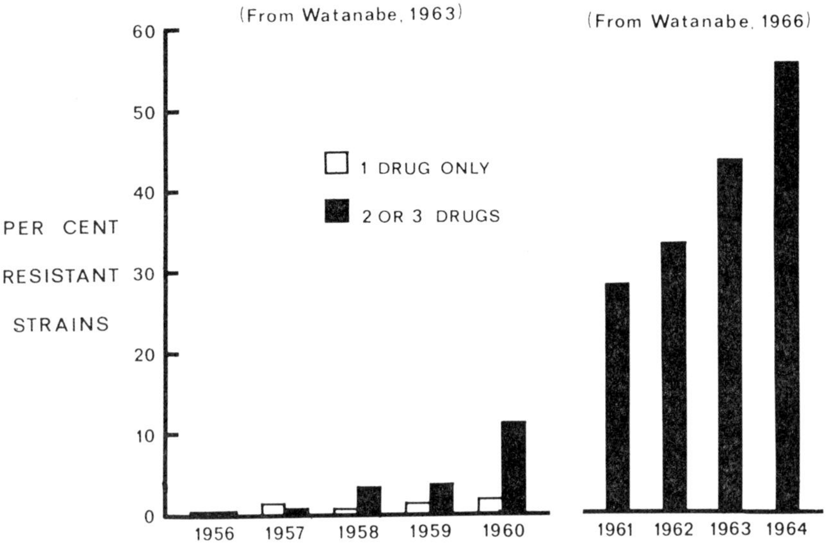

FIG. 3:7 Emergence of drug-resistant shigellae in Japan

changing social organisation will change the types of infection from which man will suffer and their response to antibiotic therapy.

1 K. W. Edmondson, 'The Growing Use of Antibiotics', *Med.J.Aust.*, **1**, 1969, 245-9.

2 R. P. Novick and M. H. Richmond, 'Nature and Interactions of the Genetic Elements Governing Penicillinase Synthesis in *Staphylococcus aureus*', *J.Bact.*, **90**, 1965, 467-80.

N. F. STANLEY I think most of us who have studied infectious disease would agree with Professor Fenner that social organisation may affect the nature and incidence of some infectious diseases in man. If this is so, it is only with a few recent events, and Professor Fenner has described some of them, that a precise relationship can be established. Today, we can begin to see some of the effects of urbanisation and fast transport between communities at the same time as we can observe some rapidly dwindling primitive societies.

In addition to Professor Fenner's 'guesses' with which I agree, I have some of my own, and I like to guess that an infection best surviving in palaeolithic man would be one transmitted vertically, or one with an incubation period commensurate with man's generation time. Do the older palaeolithic agents still remain in a modified form? The conversion of measles from a zoonosis to a human pathogen is a likely possibility and illustrates an important principle; but some consideration should be given to the possibility that its acute form arose from an older, slow form and that the slow form remains today as subacute sclerosing panencephalitis.

Not only might a slow or attenuated infection become acute or virulent, but a latent infection might become manifest in a new strain of host. There is a classic modern example of this—the first occurrence in Icelandic sheep of visna, maedi, infectious pulmonary adenomatosis, and Johne's disease in 1935, two years after the introduction (after many centuries of isolation) of Karakul sheep from Germany. The Karakul sheep were quite healthy. More subtle socio-environmental changes might have the same effect.

Take for example kuru.

'In the nearly 12 years of investigation, the clinical and pathological features of kuru, defining it as a sub-acute fatal progressive degenerative disease of the central nervous system, have remained constant; and the disease remains confined to the Fore linguistic group of the Eastern Highlands [of Papua-New Guinea] and the neighbours with whom they intermarry. Dramatic changes, however, have taken place in the sex and age incidence of the disease, with virtual disappearance of the disease among children. This has probably been the direct result of socio-environmental changes taking place in the kuru region since 1957. In the laboratory, the disease has been transmitted from a number of cases to chimpanzees after an incubation period of two years; no other host, including 17 different species of primates, has so far proved susceptible.

The most economical hypothesis that accommodates these findings without stretching the other facts, positive and negative, known about the disease is a conditionally infective agent (slow virus) which demands the proper genetic and socio-environmental background for its transmission. The sociological phenomenon best able to explain the sex and age distribution of kuru, its historical spread and its more recent changing pattern is cannibalism.'[1]

An evolutionary principle which should be emphasised is that of adaptability, demonstrated so well by man himself. The evolutionary success of influenza is dependent on its capacity for antigenic change, rather than the number of its serotypes. The development of resistance to antibiotics in bacteria, as discussed by Professor Rubbo, and the importance of episomes is another example. Our capacity to control such disease in the face of air travel and the increase of the effective population pool to world-wide proportions is going to depend not on the static complexity of an infectious agent, but on its dynamic capacity for adaptation, either in its antigenic or its metabolic attributes.

As man passed through the neolithic revolution, there always remained some who maintained the old way of life. This final link with our past will soon be broken and we are the last generation of mankind who will have the opportunity of recording *from direct observation* the way of life, social organisation, beliefs, technology, and diseases of primitive human societies. Australia must then, at this moment, demonstrate some further responsibility with regard to its Aborigines and New Guinea natives.

Arising from this discussion there are some vitally important practical steps that can be and should be taken to assist investigators of the future. With infectious disease it is obvious that sera and tissues should be frozen, and cells preserved in liquid nitrogen at the very least. Stocks could well be made from each of the Australian capital cities, as well as from Aborigines and other primitive societies. This, however, is but a part of a more formidable requirement which I hope will be discussed later.

1 Quoted from M. Alpers, 'Kuru, Some Reflections', 1968 (personal communication).

Discussion

CLEMENTS Several references have been made today to measles and the possibility of various ethnic groups being more susceptible or less susceptible than others to sudden introduction of measles and we have read in various textbooks and literature about Pacific Island groups being decimated by the introduction of measles. I have been wondering since my experience three years ago whether this is, in fact, the explanation. I was fortunate, or unfortunate, to be in the Northern Territory on the occasion of the last epidemic of measles, when something like 4,000 Aborigines north of Katherine contracted measles, in fact everybody or practically everybody under the age of fifteen was infected, and there were four deaths. I am not a virologist, but I believe it was a fairly intense infection because the rashes were confluent and I saw something like seven hundred cases and practically every one of them had massive lesions in the mouth, and the children, particularly the young children, were very sick. Now the fact that there were so few deaths may well have been due to the application of modern medical techniques and practices. I believe dehydration to be an important factor in the effects of infectious diseases, particularly in climates where the atmospheric temperature is around 90°F., and this was the temperature there at the time. Left to themselves Aboriginal mothers will not give their children fluid. We collected together all the sick people into one area in each settlement and we introduced a water patrol whereby every half-an-hour children got six, eight, or ten ounces of water. Anybody who got a respiratory infection was placed on antibiotics and given postural drainage. And I just wonder, this is the question I put to Professor Fenner and Professor Stanley, whether this question of modern understanding of perhaps the mechanisms of death in infectious diseases of this kind might not explain the absence of a high mortality where you would have expected it amongst the Aborigines and the presence of a high mortality in Pacific Island groups prior to, or in the absence of adequate modern medical methods.

FENNER I am sure that what Dr Clements suggests is in part quite correct. The lower mortality that was experienced in this outbreak of measles was undoubtedly influenced by the application of adequate modern medical methods. The fact that most children had very severe rashes is probably indicative of a lower genetic resistance of Aborigines than of white children. You said that the rashes were severe—severe, that is, in relation to what you would expect to see in white children, but I am sure the virus is one that commonly circulates in Australia. It is clear that in the epidemics of measles in the Pacific Islands, when measles was a new disease, the high death rate was due in part at least to the fact that the whole population was susceptible, not only the young children. The whole of a village population contracted the disease within a brief period of time, which meant that collecting and cooking food, looking after the sick, etc. were done inefficiently so that there was severe social disruption. The extent to which concurrent or supervening bacterial infection, which can now be controlled with antibiotics, is important in determining lethality must vary from one occasion to another. It used to be thought that secondary bacterial infection was the major, or a major, factor in deaths from smallpox; but it is now clear that smallpox can be just as lethal in the presence of antibiotic therapy as in its absence—viruses can be lethal and viruses plus streptococci can be more lethal.

FREEMAN May I make a minor, but I think not unimportant point? A phrase that has crept into the discussion both this morning and this afternoon is 'neolithic revolution'. This term was introduced into scholarly discourse by Gordon Childe, a prehistorian with explicitly Marxist leanings. We now know that the so-called 'neolithic revolution' extended over several millennia and was preceded by the Mesolithic. In other words, it was a long-drawn-out transitional period (covering a hundred and more generations) within which selection may well have occurred. I would suggest then that the term 'neolithic *revolution*' has no place in the vocabulary of an evolutionary thinker, and might well be dropped from the proceedings of this symposium.

BROWNLEA Professor Fenner—you have mentioned the sizes of communities. Have you looked at the rates at which these communities have added to their sizes in urban conditions? For example, the Wollongong/Port Kembla area is adding to its population at 7, 8, 9 per cent per annum and in notifications of infectious hepatitis the summer seasonality has now changed to an all-year incidence. I am wondering whether or not the rate at which urbanisation takes place affects the seasonality of some viruses.

FENNER I did not consider the sort of problem you present because I was sketching a very broad canvas that covered the whole evolution of mankind and all the infectious diseases he is heir to. The factor you mentioned would be important if the migrants consisted in the main part of susceptibles. If they came from other city areas where they already would probably have become immune to hepatitis, for example, one would not expect a high migration rate of (resistant) individuals to play any role, although amongst any group of migrants there must be a proportion of young and susceptible individuals.

There may also be, and this was certainly very important in true urbanisation, the recruitment of rural populations to the cities. The important fact about that is that the rural populations were almost universally susceptible to almost all the infectious diseases of the great cities, so that one was enlarging the city's population with large numbers of highly susceptible individuals. In the modern situation this factor is much less pronounced, but it may be enough to change the seasonality of a disease like hepatitis.

MIMS When you introduce new people into a city or into an area, there are big changes in those people. There has been a lot of work with animals on the effects of introducing new individuals into existing societies—and it is known there can be very great changes in their susceptibility to disease, perhaps by way of an adrenal cortical response and other mechanisms. So much so that Christian recently published a paper suggesting that some of the classical epidemiological work of people like Topley, Wilson, Greenwood, and Webster may have to be looked at again.[1] The foreign mouse or person coming into the community may differ from the others because he lacks immunity to a disease. But he also differs because, socially, he is an outsider, or, in the case of mice, a socially subordinate mouse, and this may have a real effect on susceptibility to disease.

[1] J. J. Christian, 'The Potential Role of the Adrenal Cortex as Affected by Social Rank and Population Density on Experimental Epidemics', *Am.J.Epid.*, **87**, 1968, 255-64.

BOYDEN I would like to ask a question in connection with your statement, or your conclusion, that most of the virus infections which affect man today are new, in the sense that they did not exist before civilisation. It seems to me that this suggestion—the possibility that the number of these agents is increasing—has very important social implications. I am not sure what is the total number of separate viruses now recognised which cause mild infections of the alimentary and respiratory tracts in man, but it must be somewhere between one hundred and two hundred now, is it not? Presumably none of these one hundred or two hundred existed, at least not in human beings, ten or twelve thousand years ago. Now the possibility that this number is increasing seems to me, as I say, to have important social implications. Each infection may by itself be relatively mild and only last a few days, but we do not know what the effects may be on the human organism of numerous repeated mild virus infections. And what may be the effect on society, when you have a situation eventually when most of the people are infected with some sort of virus most of the time? The question I would like to ask is whether you would hazard a guess, on the basis of existing knowledge, as to the *rate* at which the number of virus diseases affecting man is increasing?

FENNER I do not think I would make a guess as to the exact number or rate. One can say, firstly, with respect to the recognition of different serotypes of rhinoviruses (the group of little spherical viruses that cause common colds), for example, that the discovery of methods for isolating these was very recent. In part the build-up from two or three up to eighty serotypes is, therefore, a matter of sampling. There are probably many other rhinoviruses that we have not yet found, perhaps in parts of the world or in populations that have not been adequately looked at. In addition, I feel pretty certain that we are going to see the evolution of 'new' rhinoviruses and we may see the extinction of some. It may be that some viruses lose the opportunity to transfer to a susceptible individual and actually die out. The only way this can be determined is by sampling, cataloguing, and storing viruses and seeing what happens to the pattern over a period of time. One important implication of Dr Boyden's comment that I did not mention arises from the recognition that poliomyelitis is essentially a relatively new disease; it was only recognised as an epidemic disease round about the end of the nineteenth century. The polioviruses are essentially enteroviruses which have differentiated in terms of their pathogenic potential. It may well be that this sort of mutational change will occur in other viruses unrelated to polio that currently infect the human intestinal tract; a mutation may occur which is irrelevant to the survival of the virus, because all it has to do to survive is to be excreted and swallowed by somebody else. But it may be very relevant to the health of the human population if it becomes generalised in the human body and localises in the nervous system as poliovirus does. This is an ever-present possibility. I am sure we will witness the appearance of antigenically novel viruses in the next fifty years, and I do not mean only influenza viruses; possibly we may also witness the appearance of viruses of novel pathogenic potential.

DUBOS (*Chairman*) Just to end on an optimistic note. While Professor Fenner was speaking I was wondering whether there was not some possibility that increasing urbanisation and the great mobility of our population would make us all part of the same community and thus eventually establish a kind of herd immunity to the infectious agents that we carry about. But I am not going to ask Professor Fenner because he is in a gloomy mood and he would say that this is not possible.

S. Bryan Furnass

Changes in Non-Infectious Diseases Associated with the Processes of Civilisation

DEFINITIONS

Disease

The 'absence of ease' implicit in the word disease stresses the subjective element of human suffering, drawing attention to the need to consider the effect of illness on the quality rather than on the quantity of people's lives. On the other hand objective assessments of community health are more often attempted by using disease in the sense of applied pathology, namely 'a condition of the body or of some part of an organ of the body in which its functions are disturbed or deranged'. Statistical descriptions of disease and causes of death are clearly open to considerable inaccuracies, both with regard to errors of diagnosis and nomenclature. Improvements in diagnostic techniques coupled with the introduction of the international classification of the seventeen principal causes of death have helped to diminish some of these difficulties over the past two decades.

Civilisation

Civilisation is defined in the Shorter Oxford Dictionary as assimilation of the Common Law to the Civil Law. From the biological point of view a more useful meaning may be deduced from the word to civilise which means 'to bring out of a state of barbarism, to instruct in the arts of life; to enlighten and refine'.

The processes of civilisation are currently more often identified with man's use of tools, associated with his progression from palaeolithic hunting and foodgathering tribes through peasant farming communities to the explosive development of industrialised urban societies. The civilisations of ancient Egypt, Greece, and Rome depended for their growth and survival on the widespread employment of human muscle as slave labour, machines playing a relatively subsidiary role in moulding the environment. By way of contrast, modern man since the early years of the industrial revolution has progressively learnt to enslave the energy latent in matter to the extent that the need for physical labour has been reduced to a minimum. As a result of the ensuing rise in economic standards, diseases associated with toiling for food and coping with natural disasters have been supplanted by those which spring from the complexities of our technological age, in which man's connection with the soil has become less direct.

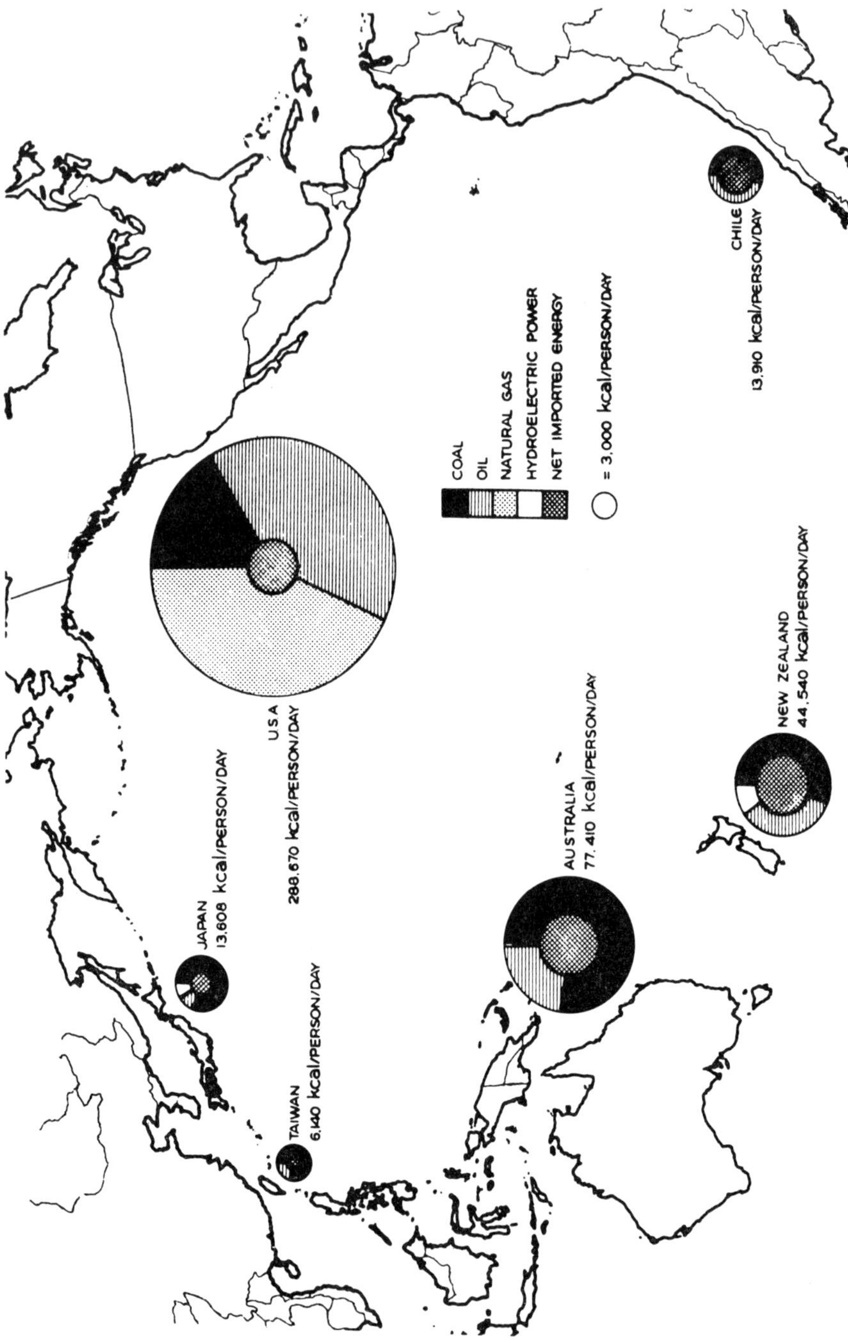

Fig. 4:1 Utilisation of coal, oil, natural gas, and hydroelectric power, selected Pacific countries, 1953-5

The relative development of technology in the Pacific area can be assessed from examination of fuel energy consumption by selected countries (Fig. 4:1). Thus, in 1955 in the United States the average citizen consumed about one hundred times as much fuel as the energy required for his personal metabolism and about fifty times as much fuel as the average inhabitant of Taiwan. It seems likely that the gap has widened over the past thirteen years since these figures were calculated.

POPULATION CHANGES

The exponential curve of world population increase (Fig. 4:2) has been associated in time with the equally explosive growth of knowledge in the physical and biological sciences. Population change has

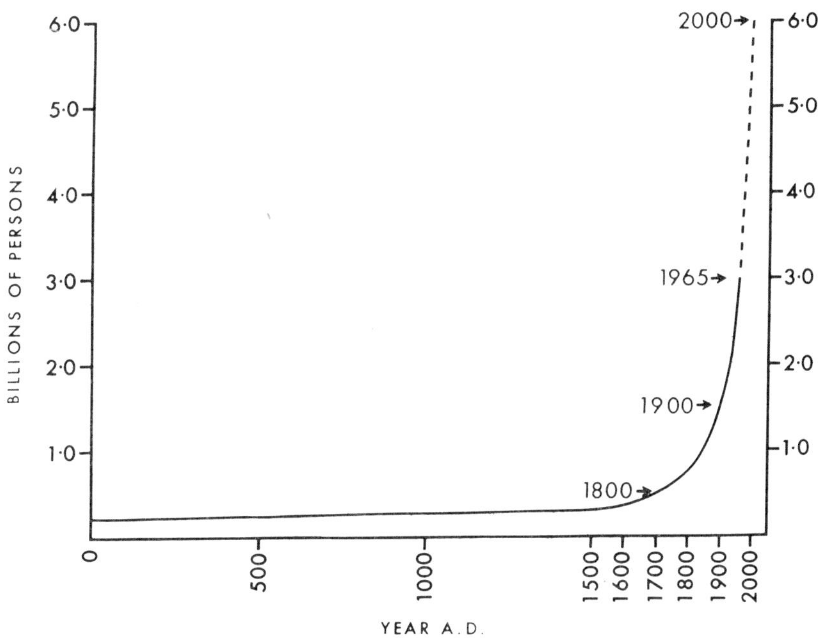

Fig. 4:2 World population

been most rapid in so-called developing countries where birth rates have remained unchanged despite a considerable reduction in death rates; an exception is found in Japan, where the fall in birth rate may have been influenced by an intensive birth control program and by legalisation of abortion (Table 4:1). Reasons which have been proposed for the fall in mortality rates include such public health measures as malaria control in post-war Ceylon and Mauritius.[1] On the other hand, cogent arguments have been advanced to support the view that falls in mortality rates are more closely related to improvements in living standards and particularly in nutrition.[2]

Table 4:1

Changes in birth and death rates in selected countries

Country	Birth rates per 1000 1940	Birth rates per 1000 1960	Death rates per 1000 1940	Death rates per 1000 1960
Mexico	44·3	45·0	23·2	11·4
Chile	33·4	35·4	21·6	11·9
Ceylon	35·8	37·0	20·6	9·1
Japan	29·4	17·2	16·8	7·6
Australia	20·3	22·6	9·3	8·5

Less often considered is the rapid post-war rate of population growth which has occurred in many industrialised countries, accompanied by a disproportionate consumption of the world's non-renewable natural resources.[3] Although the effects of the post-war 'baby boom' have slackened off, the rates of natural increase in population among industrialised nations remain highest in Australia and the United States of America (Table 4:2), the population of the latter having risen by no less than 43 per cent, from 140 millions to 200 millions over the past two decades.

Table 4:2

Rates of natural increase of population (annual excess of births over deaths per 1000 population, selected industrialised countries)

Year	Australia	United States	United Kingdom	France	West Germany	Sweden	Japan	World (estimated)
1947	13·4	15·7	8·3	8·2	4·9	8·1	19·7	—
1966	10·3	9·0	6·1	6·8	6·5	5·8	6·9	18

Calculated from United Nations Demographic Yearbook, 1967.

The absolute increase in numbers of people living in industrialised communities has been accompanied by a migration of the labour force from rural areas into the cities. Australia, with its vast land area, is now paradoxically one of the most urbanised countries in the world. At the turn of the century, inhabitants of metropolitan areas accounted for 37 per cent of the population compared with 58 per cent at the 1966 census.[4] The drift to the towns has been most marked in the post-war period, the proportion of urban inhabitants having risen from 69 per cent to 83 per cent of the total between 1947 and 1966. Between 1947 and 1954 there was a reduction of nearly half a million people living in rural areas despite an increase of 1·4 million in total population, a phenomenon which must be partly attributable to the mechanisation of agriculture. That the attractions of city life continue to provide an added incentive to migrate is suggested by Borrie and Spencer[5] who estimated that in 1961 nearly a quarter of a million persons aged seven and over residing in urban areas had emigrated from rural areas during the previous five years, and that the majority of these were young people fresh from school in the young adult age groups. Table 4:3 shows that this tendency towards centralisation continues in Australia.

Table 4:3

1966 Australian census

	Population millions	% of total	Intercensal change c.f. 1961 %
Metropolitan	6·72	58	+15·5
Other urban	2·90	25	+7·9
Rural	1·92	17	—3·3
Total	11·54	100	+9·9

Despite the obvious advantages of the city for human development, the urban sprawl which now afflicts much of Western civilisation poses increasing physical problems of road congestion and environmental pollution, together with the social disadvantages associated with migration, poverty, and loss of interaction with the natural environment.

MORTALITY RATES

Causes of death have held a fascination for man for a long time. Regular record keeping in England began in the time of Henry VIII, when the vicar of every parish was enjoined to keep a true and exact register of all weddings, christenings, and burials. From 1603 Bills of Mortality for London were issued as a continuous weekly series by the Company of Parish Clerks.[6] The Bills indicated the numbers of persons of each sex who had died from particular causes, and also the numbers in each age group; but the two factors, age and cause of death, were not interrelated until 1838, when the General Register Office was established in London.

Woodruff[7] has traced the mortality records for South Australia since records were started in 1842. There has been a remarkable and continuous change in the proportional mortality ratio, due to a decline in the death rate of infants and young children. In 1845 few more than 5 per cent of all deaths were in persons over fifty years; at the turn of the century this proportion had risen to 40 per cent for both sexes, while in 1963 80 per cent of all deaths in males and 86 per cent of all deaths in females occurred in the over fifty age group.

The changing mortality pattern for Australia as a whole may be seen by examining the age-specific death rates based on population estimates from the census years 1911 to 1966. Fig. 4:3 illustrates the remarkable decline in death rates in the newborn, which now stand at little more than a quarter of the rates for 1911. Between 1911 and

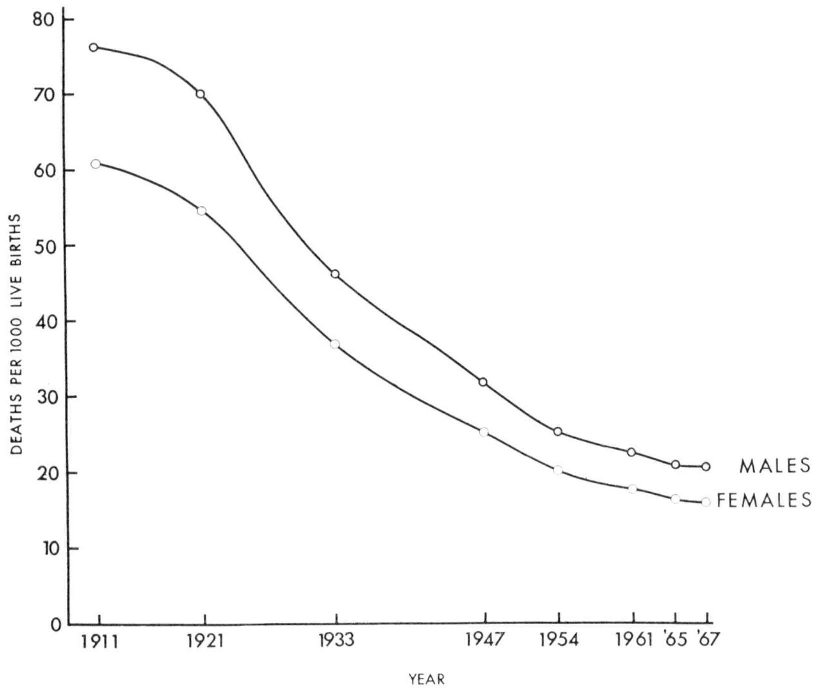

FIG. 4:3 Infant mortality in Australia

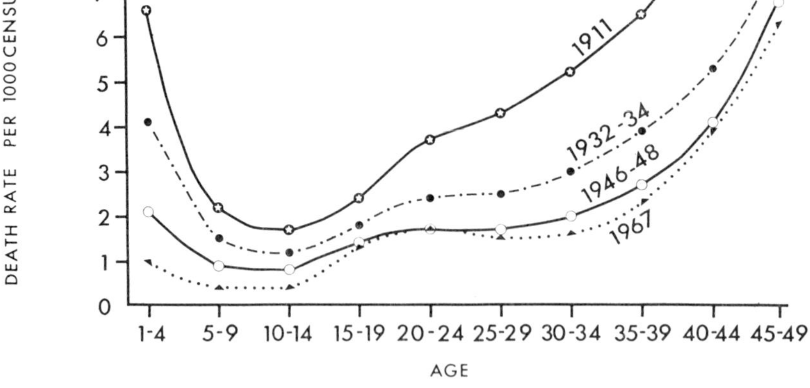

FIG. 4:4 Age-specific death rates in Australian males, 1911-67. Ages 1-4 to 45-49.

1932-4 there was also a substantial fall in death rates in persons aged one to forty-nine (Fig. 4:4), the improved outlook of young adult males in particular being due to a regression of 'the captain of the men of death'—tuberculosis. Between the immediate post-war period and 1967 the changes in male death rates appear unimpressive, particularly when seen in the context of the dramatic and well-publicised developments in curative medicine and surgery which occurred during this time. One of the reasons for this failure of further improvement is a change in causes of mortality from diseases which are now curable to those in which treatment has little to offer. For example, although there were approximately 1,300 fewer deaths from pulmonary tuberculosis in 1967 than there had been in 1950 this was more than offset by an increase in over 2,000 deaths from malignant disease of the lung. For the age groups fifty to sixty-nine there has been a steady decline in female death rates over the past thirty-five years but no corresponding improvement in male mortality (Table 4:4), a phenomenon which is largely attributable to the rising death toll from cardiovascular disease.

Table 4:4

Age-specific death rates in Australia

Year	Age group 50-54		Age group 55-59		Age group 60-64		Age group 65-69	
	Male	Female	Male	Female	Male	Female	Male	Female
1911	15·4	10·8	21·7	15·2	31·5	21·8	48·3	36·0
1932-34	11·4	8·6	17·9	15·5	26·3	17·8	40·0	29·2
1946-48	11·4	7·6	17·7	10·7	27·5	16·5	42·4	26·3
1967	10·1	5·9	17·1	8·6	27·3	13·5	43·7	21·1

The overall lowering of mortality which has occurred in response to changing conditions of living in Australia during the present century is reflected in the improved expectation of life for different age groups (Fig. 4:5), obtained from Australian Life Tables.[8] The greater improvement in expectation of life of females relative to males has led to an increase in the numbers of aging widows and fatherless children in the community, raising new psychological and social problems for solution.

Changes in immediate causes of death

The most striking change in mortality statistics over the past half-century has been the virtual disappearance of infectious disease as a cause of death in Western communities. Australian death tables for 1911 attributed 30 per cent of all deaths to infectious disease, mainly from tuberculosis and bowel infections, while an additional 5 per cent of deaths in adults were attributed to pneumonia.[9]

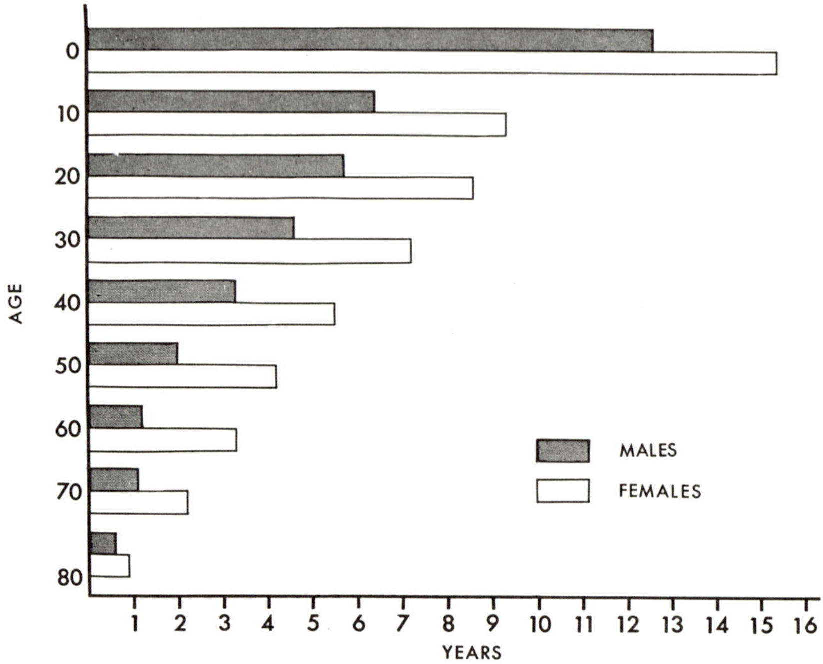

FIG. 4:5 Gain in years of expectation of life in Australia, 1960-2 compared with 1901-2; selected ages

In technologically developed nations which enjoy a high standard of material comfort the survival of increasing numbers of persons into middle life as a result of improved standards of nutrition, hygiene, and obstetric care has had the effect of increasing the proportion of deaths which are attributable to the so-called degenerative diseases. The pathological features which are characteristic of these disorders are of two main types: the first comprises neoplastic change which can lead to death from inanition, secondary infection, or pressure on vital organs; the second is characterised by obstruction within the circulatory or respiratory systems along which oxygen and nutrients are transported to the tissues from the external environment.

The main causes of death today in Australia show substantially the same trends as in other industrialised countries although there are important geographical differences. Cardiovascular disease and cancer account for two-thirds of the total mortality in both sexes (in Japan for 50 per cent). In men coronary heart disease accounts for two-thirds of mortality from all cardiovascular diseases. Lung cancer accounts in general for between 20 and 40 per cent of deaths from malignant disease; the highest rate is in Britain, while low rates are found in Norway, Sweden, and Japan. The proportion of stomach cancer deaths varies from 10 per cent in the United States to 23 per cent in Finland and 50 per cent in Japan, although the rates are tending to decline, except in Japan.[10]

THE LOST YEARS

Although mortality is inevitable, deaths which occur in old age do not as a rule engender the same sense of personal, social, or national deprivation as those which occur in the prime of life. One way of assessing the significance of premature death is to calculate the lost years of working life in a population by multiplying the numbers of deaths in each age group from fifteen to sixty-four years by sixty-five less the mean age of the group and summating the products. This type of computation has been made by Snow[11] for Western Australia and by Wells and Kupkee[12] for the Australian population as a whole.

Table 4:5 compares calculation of the lost years of working life for

Table 4:5

Changes in main causes of lost years of working life in Australian males

	1950		1967		
	% of deaths	Lost working years per 1000 population at risk	% of deaths	Lost working years per 1000 population at risk	Change in lost years
XVII Accidents, poisonings, violence	16·6	26·8	19·7	30·9	+4·1 (15%)
VII Diseases of circulatory system	34·2	19·8	40·7	22·2	+2·5 (12%)
II Malignant neoplasms	14·0	9·6	17·3	11·2	+1·6 (16%)
VI Diseases of nervous system	9·9	6·9	7·6	4·9	—2·0 (30%)
I Infective and parasitic diseases	6·5	6·5	0·8	0·6	—5·9 (90%)
IX Diseases of digestive system	5·2	4·2	3·1	2·4	—1·8 (40%)
X Diseases of genito-urinary system	5·2	3·6	1·7	1·2	—2·4 (66%)
VIII Diseases of respiratory system	5·1	3·5	4·6	2·7	—0·8 (23%)
Remaining nine causes	3·3	3·8	4·5	4·0	+0·2 (0·5%)
All causes	100	84·7	100	80·1	—4·5 (5%)
Non-infectious diseases	93·5	78·3	99·2	79·6	+1·6 (2%)

Australian males in 1950 and 1967, based on the age-specific death rates for the six most important causes of death. The emphasis placed on premature death in this type of calculation is seen by the fact that whereas the death rate from accidents, poisoning, and violence is only half that for diseases of the circulatory system, the number of lost years of working life is about 40 per cent greater.

The numbers of lost years of working life for the two most important causes of death, from accidents, poisonings, and violence (I.C.D. XVII) and from diseases of the circulatory system (I.C.D. VII) are plotted in Fig. 4:6. In 1967 at the peak death rate between ages fifteen

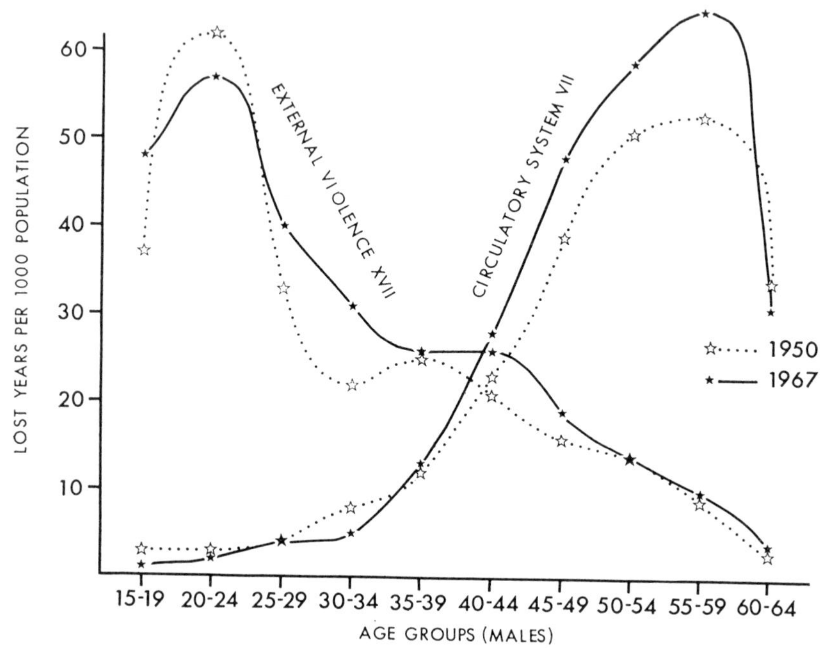

FIG. 4:6 Lost years of working life from specific causes XVII and VII

and twenty-four, motor vehicle accidents accounted for approximately two-thirds of deaths in category XVII, while at the peak between ages fifty and fifty-nine, coronary heart disease accounted for 90 per cent of deaths in category VII. Between 1950 and 1967, the number of lost years of working life from diseases of the circulatory system has increased by 12 per cent. Over the same period the number of lost years of working life from accidents, poisoning, and violence has increased by 15 per cent, partly as a result of a rise in the number of motor vehicle accidents and partly because of an increasing suicide rate.

The third commonest cause of lost years of working life in Australia is from malignant neoplasms. Fig. 4:7 depicts the death rates from malignant neoplasms between 1950 and 1967. These show a general decline of death rates except for cancer of the lung which in males shows an alarming nearly threefold increase, so that in 1967 it accounted for 27 per cent of all deaths from malignancy in the male compared with 11 per cent in 1950. The changing lung cancer rate which is matched by a rising death toll from chronic bronchitis increases more steeply in the older age groups.

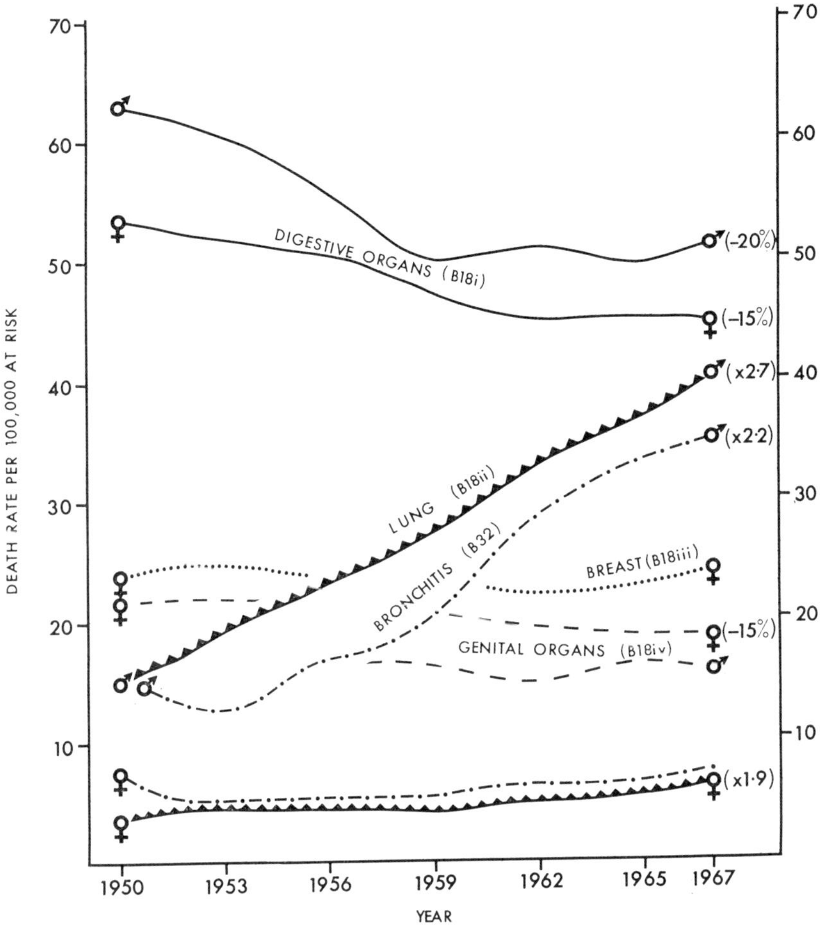

FIG. 4:7 Australian death rates for principal malignant diseases and bronchitis (the figures in parentheses which follow the names of diseases are derived from the International Code of Classification)

MORBIDITY SURVEYS

Mortality statistics are clearly only crude measures of assessment of causes of death and can give little guide to the amount of illness (morbidity) in the community at large. Studies of morbidity in general practice have been carried out in several countries including England and Australia.

An English survey[13] has indicated that approximately 30 per cent of the general practitioner's patients suffer from chronic disease, most of which is irreversible, notably cardiovascular, respiratory, and rheumatic disorders, which occupy more than three-quarters of the doctor's time. In Australia, a community health survey in the small country town of Heyfield, Victoria[14] found evidence of disease in 60 per cent of the population, the incidence being lowest in children (34 per cent) and highest amongst the elderly (89 per cent). Among medical disorders, a high incidence of asthma, hay fever, and chronic bronchitis was found.

Mental illness

Lack of agreement on what constitutes mental illness and how it may be measured creates many difficulties in assessment of its true incidence in the population. The size of the public health problem is not generally realised. In 1960, 48 per cent of all hospital beds in England and Wales and 44 per cent of beds in the United States were occupied by patients who were mentally sick or mentally defective—about one occupied bed per 300 total population.[15] However striking these figures, they do not give an adequate picture of the amount of mental sickness in the community, for the greater part of this sickness is treated in outpatient clinics or by general practitioners.

In Australia, a morbidity survey carried out in 1962-3 by eighty-five general practitioners indicated that 3·3 per cent of all episodes of illness commencing during the year were for mental, psychoneurotic, and personality disorders, 61 per cent of these episodes occurring in females.[16] The Heyfield community survey suggested that some degree of psychiatric disability was present in approximately 18 per cent of the population, the majority of whom were not receiving any form of psychiatric help.

Apart from their value in defining the incidence of disease, community health surveys clearly serve as useful models for studying some of the possibly preventable difficulties in interpersonal and social relationships which so often impair the quality of people's lives.

ENVIRONMENTAL AGENTS OF DISEASE

The argument is less often heard these days as to whether nature *or* nurture is responsible for disease; rather does the question more often arise as to what extent heredity *and* environment may be playing their respective roles in the emergence of illness in any particular instance. Inherited predisposition towards diabetes and other types of biochemical abnormality are now well recognised, as is a genetic contribution towards hypertension[17] and coronary heart disease.[18] Less frequently considered is the possible influence of the elimination of natural selection on the genetic composition of a population. An example is the unmeasured (and possibly unmeasurable) effects on the constitutional robustness of 'civilised man' which may have resulted from the early exhibition of powerful chemotherapeutic agents such as antibiotics.

Although an elaborate system of medical care existed in ancient Egypt under the leadership of Imhotep who lived about 3000 B.C., there are no known recorded thoughts concerning environmental contributions to disease in Western civilisation until the time of the Hippocratic School in Greece in the fifth century B.C. In the first recorded treatise on medical geography entitled 'Airs, Waters and Places',[19] Hippocrates describes the successful physician as one:

'who has a due regard to the seasons of the year, and the diseases which they produce; to the states of the wind peculiar to each country and the qualities of its waters; who marks carefully the localities of towns, and of the surrounding country, whether they are low or high, hot or cold, wet or dry; who, moreover, takes note of the diet and regimen of the inhabitants, and in a word, of all the causes that may produce disorder in the animal economy.'

Hippocrates's wisdom appears to have a great deal of relevance to the study of modern diseases of civilisation.

AIR POLLUTION

Man's use of the solar energy latent in wood and coal for domestic and industrial purposes and his subsequent harnessing of energy in oil products for transport has resulted in increasing atmospheric pollution by sulphur compounds and hydrocarbons. The concentration of these air pollutants will clearly be influenced by population density, concentration of industry, domestic heating and by the number of motor cars in use. If atmospheric pollution does constitute a health hazard one would expect to find, in some countries at least, a difference in mortality rates between rural areas and towns. In 1947 Stocks[20] advanced the hypothesis that 'smokiness of the atmosphere is an important factor in itself in producing cancer of the lung', on the ground that the mortality in towns was inversely related to the amount of sunshine recorded. In Britain at the time of the 1951 Census the standardised mortality ratio for males in the conurbations (outside London) was double that in the combined rural districts of England and Wales. Subsequently, Stocks[21] found that mortality for lung cancer and bronchitis was highly correlated with the amount of smoke deposited, and that of the four common hydrocarbons in polluted atmospheres, 3:4 benzopyrene was the most closely related to mortality. The hypothesis that air pollution contributes significantly to the development of cancer of the lung later received support from the observation of Dean, who reported that British male immigrants to South Africa[22] and to Australia[23] had a higher mortality rate from this disease than did the local-born inhabitants, despite the latter's higher rate of cigarette consumption.

Industrialisation in Australia must now be contributing to air pollution, an index of which is suggested by the sixfold rise in imports of crude petroleum into the country between 1947 and 1963. Although much of this increase will have been used for industrial production and in the conversion from steam to diesel power on the railways, a fair proportion of it must have been consumed in running private motor cars, the number of which has doubled over the same period. An index of how popular this mode of transport has become is revealed by the fact that only 0·4 per cent of all vehicles on the road are omnibuses.[24]

Although urban man may appear to adapt to living in a polluted environment, the progressively rising incidence of cancer of the lung in older age groups of Australian males suggests that a co-carcinogen in the air may only exert its lethal effects after a decade or two of exposure, thereby obscuring the detection of cause and effect relationships. An example of how air pollution can be rapidly and dramatically fatal was seen in the London smog of 1952 when 4,000 people died of respiratory failure during a 3-day temperature inversion.[25]

Man's proclivity for living in cities appears to carry an increased risk of developing coronary heart disease. In his National Atlas of Disease Mortality in the United Kingdom, Howe[26] showed that the mortality rate from coronary heart disease was lower in rural areas compared with urban centres of population. In the United States Berkson and co-workers[27] have reported that mortality from all cardiovascular and renal diseases is approximately twice as high in Chicago as it is in rural Illinois, while in Norway it has been demonstrated that six rural centres of population had mortality rates for arteriosclerotic and degenerative heart disease and for malignant tumours only a third of the corresponding rates in Oslo for the period 1952-61.[28] A more puzzling environmental association which would have interested Hippocrates has been revealed by epidemiological studies in several countries indicating that death rates from cardiovascular disease (and in Britain from bronchitis) are higher in areas with soft than in areas with hard drinking water.[29]

Nutrition and disease

Hippocrates's dictum that the fat die sooner than the thin finds statistical confirmation in modern Western man. In the United States, the aggregate mortality of persons who are 20 per cent or more overweight is approximately 50 per cent above the expected rate, leading causes of premature death being cardiovascular disease, hypertension, diabetes, gall bladder disease, and cirrhosis of the liver.[30] A retrospective study conducted in a London clinic in 1959 found that among 373 consecutive outpatient admissions, 101 patients were 20 per cent or more above standard weight and were obese by study definition.[31] This number represents 26 per cent of the men and 44 per cent of the women. Approximately one-third of these 101 patients had been obese from childhood and showed a high incidence of psychological abnormalities as evaluated by clinical examination.

The causes of obesity (derived from *obesus*, the past participle of the Latin *obesere*, meaning 'that which has eaten itself fat') appear to include both genetic and environmental influences. Access to unlimited and diverse food supplies, particularly when these are processed for palatability, provides incentives for over-nutrition which do not exist amongst some primitive communities (Fig. 4:8). On the other hand, several studies suggest that in some groups physical underactivity may be more important than overeating as a contributing cause of obesity.[32]

Many investigators have attempted to associate particular items of the diet of Western man with his high incidence of coronary heart disease. For example Keys[33] has linked mortality rates in different countries with consumption of animal fat, whereas Yudkin[34] believes that a high intake of refined carbohydrate, particularly sugar, is more likely to be causally related to coronary heart disease, as it is to dental

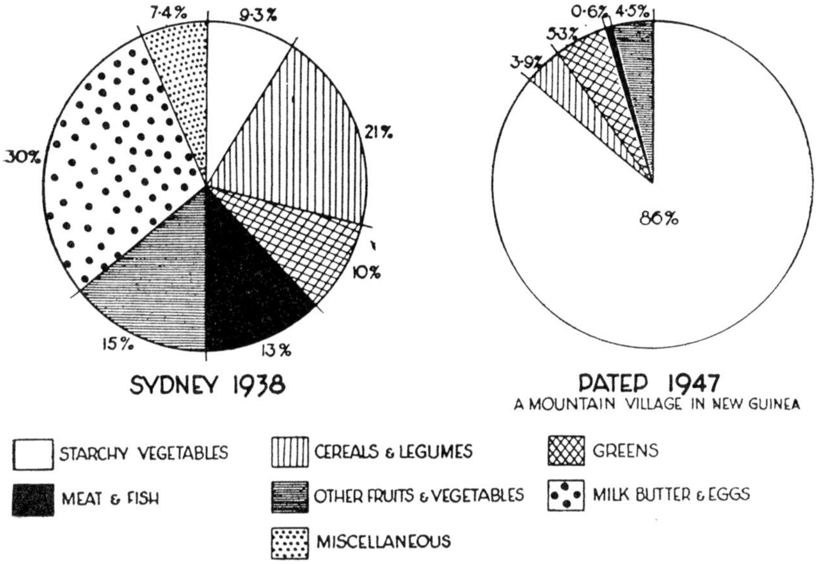

FIG. 4:8 Comparison of the percentage contribution by weight of various types of foodstuffs in the Australian and New Guinean diets

caries and obesity. On the other hand the sex difference in death rates from coronary thrombosis, coupled with the apparent associations between heart disease and other cultural influences including the pattern of physical activity, the cigarette smoking habit, and possibly mental stress, suggest a multifactorial rather than a simple dietary cause of this common affliction of industrialised societies. It is possible that obesity and coronary occlusion may represent a final common pathway resulting from rates of intake and mobilisation of energy-rich substances which are consistently in excess of rates of energy expenditure.[35]

EXERCISE AND HEALTH

A century ago it was unavoidable that almost everyone in the population frequently undertook exercise of a moderate degree. No public transport was available; to sit in a carriage and be conveyed anywhere, except for long journeys, was the prerogative of the privileged few. Walking was therefore a regular form of exercise for everyone, and considerable distances must have been covered daily by many people. Most of the population would also have been employed in work necessitating physical labour, sometimes of marked severity.

The development of public transport, first in cities and later over most of the country, together with urbanisation of the community and, in the past forty years, the motor car, the most potent influence of all, have caused a major revolution in our way of life. Technology has brought with it automation in factories and a great increase in the number of people in sedentary occupations, while the development of spectator sports and forms of passive entertainment such as television means that an increasing proportion of leisure time is spent standing or sitting. Mechanical appliances in the home, while taking much of the drudgery out of housework, have deprived the housewife of the need for muscular activity, reflected in objective measurements of her generally low levels of energy expenditure.[36]

The hypothesis that habitual physical activity protects against coronary heart disease is derived from statistical comparisons which show that incidence, prevalence, and mortality rates vary inversely with the average levels of exercise involved in different occupations. Thus Morris and his colleagues[37] have suggested that it is the extra exercise necessitated by the work of London bus conductors, in running up and down stairs, that results in a much lower incidence of coronary thrombosis than that suffered by drivers. Similarly, the considerable amount of physical activity of postmen seems the most obvious difference between them and other Post Office workers such as telephonists and clerks, who are more prone to develop heart attacks.[38] From the clinical point of view an American investigator has found that inactive men show both an increased incidence of coronary occlusion and a higher fatality rate than do active men.[39]

Occupational activity levels are of diminishing interest for public health, since energy expenditure for nearly all types of work is diminishing with technological advance. Observations of people in clerical occupations have indicated that despite the small amounts of physical energy used up in leisure, more energy is frequently spent outside the work period than at work itself.[40] The Social Medicine Research Unit in London have planned a survey of leisure activity in a large group of public servants in order to observe prospectively whether incidence rates of heart disease are less in those graded as physically more active in their leisure.[41] Some objective support in favour of physical activity has been found in a small group of university students, the most active members of the group showing significantly better tests of physical fitness and of academic performance than the least active participants.[42]

Passmore[43] has speculated that man might be divided into three hypothetical sub-species—'Homo laborans', 'Homo sedentarius', and 'Homo sportivus'. Since civilisation began, 'Homo laborans' constituted the vast majority of people in all countries until recent times and is still in the majority over most of Asia. In Western civilisation, technical developments and a century of industrial legislation have resulted in considerable reduction in hours of work to a point when a twenty-hour week seems feasible for most of the population of industrialised societies, within a few decades. While the resulting emergence of 'Homo sedentarius' may indicate the short-term temporary adaptation of man to our technological age, the rising incidence of coronary heart disease and obesity may represent a long-term frustration of normal function.

Members of the sub-species 'Homo sportivus' need no justification for their attitudes since they enjoy participation in physical activity, either in competitive sport or from such positive interaction with their natural environment entailed in walking, climbing, swimming, fishing, sailing, skiing, or dancing. Since the evidence suggests that a moderate degree of muscular activity can not only enhance the enjoyment of living but may also exert a long-term adaptive effect which is beneficial to health, it would seem only sensible for authorities, at all levels from government downwards, to provide education and facilities for suitable forms of active leisure. Perhaps 'Homo ludens' (man at play) would be a more suitable description than 'Homo sportivus', including as it does other types of human creativity involved in the fine arts, performing arts and craftsmanship which compose the hallmarks of civilised man.

Although the processes of civilisation inevitably commit man to living in cities, the conurbations of increasing area and density which he now inhabits deprive him of contact with his natural environment to an extent which did not exist in mediaeval cities. This danger is being recognised in North America, where the preservation of wilderness for recreational purposes is regarded as a vital aspect of mental health.[44]

Drugs

From time immemorial man has attempted to amend the harshness of his situation and diminish his sense of isolation by selecting for consumption certain chemicals in his environment which change perception. The use of the South American plant peyotl by the Aztecs is an example of the then socially accepted method of altering the state of consciousness in religious ceremonies. In seeking solace for his problems, modern Western man has access to a vast array of plant and chemical substances which are manufactured for profit and which provide a useful source of revenue for governments. Affluence has placed the consumption of these substances within the reach of almost everyone in industrialised countries, including the adolescent age group.

Addiction to nicotine (named after Jaques Nicot who introduced tobacco into France in 1560) provides the best documented example of the association between a faulty personal habit and disease. In 1957 the Medical Research Council advised the British Government of:

'a very great increase during the past 25 years in the death rate from lung cancer in Great Britain and other countries. Evidence from many investigations in different countries indicates that a major part of the increase is associated with tobacco smoking, particularly in the form of cigarettes.'[45]

Ten years later a World Conference on Smoking and Health[46] supported these conclusions and pointed out that cigarette smoking also contributed substantially to the likelihood of death or disability from coronary heart disease and from chronic bronchitis and emphysema, environmental pollution providing an additional hazard in the latter instance. In some countries, cigarette smoking may make a greater contribution to death from coronary heart disease than from chronic respiratory disease. Hammond[47] has described epidemiological studies embracing almost half a million male Americans over the period 1959 to 1965. He found, *inter alia,* that about three-quarters of non-smokers aged 25 might be expected to live to the age of 65 years, but only about half of continuing smokers of forty or more cigarettes per day might live to that age, nearly half the excess deaths being attributable to heart disease and only 14 per cent to lung cancer. Recognition of cigarette smoking as a major environmental hazard to Australians has appeared in a recent publication of the National Heart Foundation.[48]

Statistics presented at the World Conference on Smoking and Health indicated that among twenty of the nations represented available data showed an average consumption of 2,400 cigarettes per year *per capita* of the total population aged fifteen and over. In the USA, forty-nine million men and women (42 per cent of the adult population) smoked cigarettes, 50 per cent of adolescents were smoking by the age of eighteen years and one million young people started smoking each year. Australian statistics for the apparent consumption of tobacco and cigarettes show a 50 per cent increase in total tobacco consumption but a fourfold increase in the consumption of manufactured cigarettes compared with pre-war.[49] Encouragement to those who wish to stop smoking comes from the finding of Doll and Bradford Hill[50] that cessation of smoking by British doctors was followed by a fall in their mortality rate, particularly from cancer of the lung.

The extent to which cigarette advertising has contributed to the increase in smoking habits is a debatable point. On the other hand in a supposedly democratic society it is hard to justify some types of cigarette advertising aimed at young people since it may interfere with their freedom of choice not to start smoking. An example is the impairment of judgment which can be induced by purporting to show a positive association between cigarette smoking and healthy living, in contrast to the increasing weight of evidence against such an association. The behavioural science which studies the link between personality types and disease is still in its infancy, but in a small group of university students, cigarette smokers showed a significantly higher score for neuroticism and extroversion than did non-smokers.[51]

Of all the chemicals acting on the central nervous system alcohol has become the most socially accepted one in Western societies. Its benefit in facilitating interpersonal communication is, however, sometimes outweighed by the social disadvantages which can follow overconsumption. Thus, in 1948 it was estimated that there were 3·8 million alcoholics in the United States out of 62 million drinkers, a ratio of 1:16.[52] A recent Australian study has estimated that alcohol is directly responsible for 3 per cent of all deaths, including 50 per cent of deaths from motor vehicle accidents and 20 per cent of deaths from suicide.[53] A Royal Commission Report in 1965 summed up the position as follows:

'On the basis of much research it seems a well-founded conclusion, and the conclusion applies to Australia, that a majority of the road accidents involving death or serious bodily injury occur because the driver or drivers involved is, or are, suffering from impairment of necessary skills due to the consumption of alcoholic liquor.'[54]

The fall in road deaths in Britain following the introduction of the breathalyser is an example of how modern legislation can reduce mortality. On the other hand prevention of the social tragedy of alcoholism can only come from a more comprehensive understanding of the social and interpersonal reasons why the use of alcohol should be so prevalent in modern society.

One of the features of life in modern industrial societies is the enormous quantities of pharmaceutical products consumed by the public at large, both as patent medicines and on prescription from medical practitioners. While some of these preparations undoubtedly play a part in delaying the onset of death or improving the quality of life, the majority appear to fulfil the role of the mediaeval talisman, aiming to propitiate the winds of ill fortune or to achieve a 'cure' for the stresses of modern living. Many of the pills and potions which are swallowed morning, noon, and night by the faithful have unfavourable effects on health, including for example the influence of APC powders in producing kidney disease[55] and of aspirin in causing gastric ulceration,[56] quite apart from the disastrous effects of drugs such as thalidomide.

It seems that over-consumption of pharmaceutical products, as of food, cigarettes, insecticides, and fuel oil, must be included amongst the environmental hazards to health which beset modern Western man.

A NOTE ON PREVENTION

Communicable diseases which accompanied the nutritional deficiencies and environmental squalor of the industrial revolution have largely been abolished through nineteenth-century legislation relating to working conditions in factories and mines. The major challenge for public health in industrialised communities now appears to lie in the control of chemical and radioactive effluents and also in drawing attention to the long-term biological hazards which may result from the indiscriminate use of the products of man's technology, including insecticides, on the balance of nature.[57]

From the point of view of personal health, it is in the nature of chronic degenerative disorders and psychoses that cure is often difficult or impossible since the underlying pathological process has been developing over a substantial portion of the life span of the affected individual and is often irreversible by the time that symptoms of disease appear.

Opportunities for secondary prevention may lie in the attempted correction of risk factors in disease, for example the cigarette smoking habit, overweight, hypertension and a raised level of serum lipids, all of which are associated with a greater likelihood of development of coronary occlusion.[58] Similarly, the early recognition of symptoms of depression or anxiety may enable treatment to be started before clinical evidence of more serious mental disorder develops.

A more promising though undoubtedly more difficult opportunity for medicine lies in primary prevention, namely the encouragement of sensible living habits from an early age. The adoption of such a philosophy is hampered in two main ways. Firstly, emphasis on so-called 'curative' medicine is fashionable amongst doctors and their patients, the search for a panacea or for dramatic developments in surgery always finding a ready market in the mass communication media, irrespective of their expense or irrelevance to community needs. Preoccupation with real or imagined illness does not necessarily diminish with improvement in the physical environment. For example, in their review of a new town in England, Taylor and Chave[59] were disappointed to find that, while there was a lessened incidence of psychosis, neurotic syndromes remained much as for the national average. To deal with the situation of neurotic illness they claimed that:

'What is needed appears to be a new moral climate, with a rather more robust attitude to minor deviations from normality. In such a robust society, there must be less running to the doctor with a trivial complaint and more getting on with life and making the best of it as it is.'

The second obstacle to primary prevention of disease in the 'free' societies of the Western world is an economic one. Thus, private manufacturers and governments alike tend to favour increasing consumer spending which yields immediate profits and votes at the expense of less tangible policies such as conservation of the natural environment, education, and encouragement of the arts, whose benefits to the community can only be measured on a long-term basis. The social implications of this dilemma have been summarised by Morris:

'If it is a fact that a major burden of disease arises from the successful functioning of our society, then the tasks confronting the "sanitary reformer" of today are as great as anything met by nineteenth century pioneers of urban and industrial hygiene.'[60]

Leisure

While science and technology have created many new work opportunities for highly skilled and gifted people in Western civilisation it has also highlighted the dangers of underemployment, particularly in the context of increasing population. Adaptive responses to these dangers include the fostering of waste and obsolescence in manufacturing industries and the creation of a growing bureaucracy in accordance with the 'law' of Professor Parkinson that work expands in order to fill the time available for its completion.[61]

In seeking an answer to the problems of how, with increasing automation in society, Western man is to fill his time, Professor Cohen, a psychologist from Manchester, poses the questions:

'How did Adam occupy his time in the Garden of Eden before the Fall? What precisely did he do day by day, hour by hour? . . . Suppose that the working week were to be cut down, by Government decree to 20 hours, with pay unaffected. We might predict two consequences: substantial rise in the consumption of alcoholic beverages, and an increase in gaming and gambling: alcohol as a defence against boredom, and gambling as a means of enlarging income for more conspicuous consumption.'[62]

Cohen states that he is 'not primarily occupied with the tribulations of permanent under-secretaries or bishops, who know how to use their free time' but rather 'with the millions who work by the clock and who are likely to find themselves with more and more free time on their hands'. Their predecessors in past centuries had no such problem, but they needed some release, and this was found in traditional festivities, some of them boisterous and remarkably uninhibited, but all of which can be included under the general heading of 'play'.

An American psychiatrist, Martin, lays strong emphasis on the responsibility of medicine in the approach to this problem, commenting:

'Modern medicine has a major interest in how individuals react and adapt to change in their cultural environment, and what such change contributes to health and disease. This new freedom of time confronts man with perhaps the most radical change in his evolutionary history.'[63]

In more creative participation in leisure activities may lie part of the answer to the primary prevention of not only premature death from the degenerative disorders but also of mental illness and social delinquency in the community.

A contribution towards improving the quality of people's lives might be made by a new breed of family physician, trained primarily in the behavioural sciences and orientated in outlook towards the prevention rather than the cure of disease. While recognising the importance of early detection and management of disease, his more positive approach to people and their problems might in its broadest sense resemble that of the Greek physician Eryximachus who spoke of medicine as 'the knowledge of the loves and desires of the body and how to satisfy them'.

A GLOBAL VIEW

No longer is it possible to think of the well-being of man outside the context of the survival and health of mankind. Such is implied in the etymology of the word health, derived from the Old English hale, meaning whole. The sad paradox of global disease is that while the majority of the world's population are greatly at risk from undernutrition and communicable diseases, many of the medical misfortunes which afflict so-called developed communities are the result of increasing affluence. The physicist Denis Gabor defined our problem as the human trilemma–overpopulation, the danger of nuclear (and biological) warfare, and the Age of Leisure–all of which can be attributed directly or indirectly to developments in technology.[64]

There appear to be two ways in which scientific endeavour may contribute towards a solution to this trilemma, if mankind is to have a future. Firstly there is a clear need for governments to recognise that the biological requirements of agrarian reform and fertility control should take precedence over all political and ideological differences. The second opportunity lies in the urgent need for research into the workings of the subconscious mind so that man's tendency towards intraspecific aggression, which can now have disastrous consequences, may be channelled into less destructive outlets.

In the final analysis, the survival and health of the individual and his species may depend not so much on further developments in technology as on the collective application of common sense.

Acknowledgment: I wish to acknowledge the help given by Mr K. Archer and staff of the Commonwealth Bureau of Census and Statistics, Canberra.

Summary

Improvements in nutrition and environmental hygiene which followed in the wake of the industrial revolution have led to the attainment of reproductive age by an increasing number of children, with consequent rapid growth of population. In communities with an advancing technology the increased expectation of life has been associated with a change in the main causes of death from infectious diseases to the so-called degenerative disorders.

During the past two decades in industrialised societies there is evidence of a growing death rate from coronary heart disease, cancer of the lung, bronchitis, and external violence, particularly in males. Environmental influences associated with these changes in death rates appear to include faulty patterns of social and individual behaviour, notably air and water pollution, the cigarette-smoking habit, overconsumption of rich foods and alcohol, and inadequate exercise. Hence under conditions of affluence and crowding the products of man's technology may be inducing a selective method of population control just as micro-organisms are the main agents of death under conditions of poverty.

Since land space and natural resources are finite, the alternative to further increases in natural or self-induced death rates appears to lie in the control of human fertility both in developing and industrialised communities. Where the incentive is available an enlightened policy of city planning, application of knowledge in the behavioural sciences, and genetic counselling appear to offer promise for enhancement of the quality of human life.

With improving communication systems a good case can be made out for a more fruitful interchange of information and resources between industrialised and developing communities regarding the prevention of disease and the promotion of health.

References

1 'Malaria and the Population Explosion', *Lancet*, 1, 1968, 899-900.

2 H. Frederickson, 'Determinants and Consequences of Mortality and Fertility Trends', *Pub.HlthRep.Wash.*, **81**, 1966, 715-27.

3 E. H. Hipsley, 'The Relationship between Energy Utilisation and Food Production: The Examples of New Guinea and Australia', Congress Symposium No. 1, 1966, 11th Pacific Science Congress, Tokyo.

4 1966 Census of the Commonwealth of Australia, Census Bulletin No. 9.1, Commonwealth Bureau of Census and Statistics, Canberra.

5 W. D. Borrie and Geraldine Spencer, *Australia's Population Structure and Growth*, 2nd rev. ed., Committee for Economic Development of Australia, Melbourne, 1965.

6 J. A. Scott, 'London County Council Report of the County Medical Officer of Health and Principal School Medical Officer', 1957, 165.

7 P. Woodruff, 'Changing Mortality: Triumph and Challenge in Preventive Medicine', *Med.J.Aust.*,2, 1963, 392-400.

8 S. W. Caffin, 'Australian Life Tables', Govt Printer, Canberra, 1961-2.

9 Demography Bulletin 29, 1911, 68, 1950. Figures for 1967 will appear in Causes of Death Bulletin No. 5. Commonwealth Bureau of Census and Statistics, Canberra.

10 J. H. De Haas, 'Geograpical Pathology of the Major Killing Disorders: Cancer and Cardiovascular Diseases', *Health of Mankind*, in Ciba Foundation Symposium, London, 1967, Churchill, 79-102.

11 D. J. R. Snow, 'Report of the Commissioner for Public Health, Western Australia', 1951.

12 R. Wells and L. Kupkee, 'The Lost Years: Important Causes of Death in Australia, 1963', *Med.J.Aust.*, 2, 1966, 466-8.

13 W. P. D. Logan and A. A. Cushion, 'Morbidity Statistics from General Practice; Vol. 1 (General)' London, 1958, H.M.S.O.

14 J. Krupinski *et al.*, 'A Community Health Survey of Heyfield, Victoria', *Med.J.Aust.*, 1, 1967, 1204-11.

15 P. A. P. Moran, 'Statistical Methods in Psychiatric Research', paper prepared for presentation to the Royal Statistical Society, London, in April 1969.

16 Report on a National Morbidity Survey, February 1962-January 1963, Pt I, National Health and Medical Research Council, Canberra, 1966.

17 R. Platt, 'Heredity in Hypertension', *Lancet*, 1, 1963, 899-904.

18 J. Stamler *et al.*, 'Coronary Risk Factors: Their Impact and their Therapy in the Prevention of Coronary Heart Disease', *The Medical Clinics of North America*, **50**, 1966, 229.

19 R. Dubos, 'Of Airs, Waters and Places', *Man Adapting*, New Haven, 1965, Yale University Press, 35-41.

20 P. Stocks, 'Regional and Local Differences in Cancer Death Rates', London, 1947, H.M.S.O.

21 P. Stocks, 'Cancer and Bronchitis Mortality in Relation to Atmospheric Deposit and Smoke', *Brit.med.J.*, 1, 1959, 74-9.

22 G. Dean, 'Lung Cancer Among White South Africans', *Brit.med.J.*, 2, 1959, 852-7.

23 G. Dean, 'Lung Cancer in Australia', *Med.J.Aust.*, 1, 1962, 1003-6.

24 *Year Book of Australia* 1967. Census of Motor Vehicles, 31 December 1962, Types of Vehicle, States and Territories, 458.

25 J. L. Burn, 'The Control of Air Pollution', *Recent Advances in Public Health*, London, 1959, Churchill, 287-301.

26 G. M. Howe, *National Atlas of Disease Mortality in the United Kingdom*, London, 1963, Nelson.

27 D. M. Berkson *et al.*, 'Socioeconomic Correlates of Atherosclerotic and Hypertensive Heart Diseases', *Assn.N.Y.Acad.Sci.*, **84**, 1960, 835-50.

[28] A. Jervell, K. Meyer and K. Westlund, 'Coronary Heart Disease and Serum Cholesterol in Males in Different parts of Norway', *Acta.med.Scand.*, **177**, 1965, 13-23.

[29] H. A. Schroeder, 'Relation between Mortality from Cardiovascular Disease and Treated Water Supplies', *J.Amer.med.Ass.*, **172**, 1960, 1902-8; G. Biorck, H. Bostrom, and A. Widstioen, 'On the Relationship between Water Hardness and Death Rate in Cardiovascular Diseases', *Acta.med.Scand.*, **178**, 1965, 239-52; M. D. Crawford, M. J. Gardner, and J. N. Morris 'Mortality and Hardness of Local Water Supplies', *Lancet*, **1**, 1968, 827-31.

[30] L. I. Dublin, and H. H. Marks, 'Mortality amongst Insured Overweight in Recent Years', *Tr.Assoc. Life Insur.Med.Dir. America*, **35**, 1951, 235-67.

[31] A. G. Mullins, 'The Prognosis in Juvenile Obesity', *Nutr.Rev.*, **17**, 1959, 99-100.

[32] M. L. Johnson, B. S. Burke, and J. Mayer, 'Relative Importance of Inactivity and Overeating in the Energy Balance of Obese High School Girls', *Amer.J.Clin. Nutr.*, **4**, 1956, 37-44.

[33] A. Keys and J. T. Anderson, 'The Relationship of the Diet to the Development of Atherosclerosis in Man', In Symposium on Atherosclerosis, *U.S. Nat.Acad.Sc.*, **338**, 1954, 181-204.

[34] J. Yudkin, 'Nutrition and Palatability with Special Reference to Obesity, Myocardial Infarction and Other Diseases of Civilisation', *Lancet*, **1**, 1963, 1335-8.

[35] E. H. Hipsley and S. B. Furnass, 'Coronary Heart Disease: Some Reflections on its Causation and Prevention', *Med.J.Aust.*, **2**, 1966, 607-21.

[36] J. V. G. A. Durnin, E. C. Blake, and J. M. Brockway, 'The Energy Expenditure and Food Intake of Middle-Aged Glasgow Housewives and their Adult Daughters', *Brit.J.Nutr.*, **11**, 1957, 85-94.

[37] J. N. Morris *et al.*, 'Coronary Heart Disease and Physical Activity of Work', *Lancet*, **2**, 1953, 1053-7.

[38] J. N. Morris *et al.*, 'Coronary Heart Disease and Physical Activity of Work', *Lancet*, **2**, 1953, 1111-20.

[39] C. W. Frank *et al.*, 'Myocardial Infarction in Men: Role of Physical Activity and Smoking in Incidence and Mortality', *J.Amer.med.Ass.*, **198**, 1966, 1241-5.

[40] R. C. Garry *et al.*, *M.R.C.Spec.Rep.Ser.*, No. 289, 1955.

[41] S. Yasin *et al.*,'Assessment of Habitual Physical Activity apart from Occupation', *Brit.J.prev.soc.Med.*, **21**, 1967, 163-9.

[42] S. B. Furnass and H. M. Whyte, 'A Student Health Survey', in press.

[43] R. Passmore, 'An Assessment of the Reports of the Second Committee on Calorie Requirements (FAO 1957)', Food and Agriculture Organization of the United Nations, Rome, 1964.

[44] L. H. Day and A. T. Day, *Too Many Americans*, Boston, 1964, Houghton Mifflin, 45-74.

[45] 'Tobacco Smoking and Cancer of the Lung' Statement by the Medical Research Council, *Brit.med.J.*, **1**, 1957, 1523-4.

[46] A. G. Maclaine, 'World Conference on Smoking and Health', *Med.J.Aust.*, **2**, 1967, 1145-8.

[47] E. C. Hammond, 'World Costs of Cigarette Smoking in Disease, Disability and Death', address to the World Conference on Smoking and Health, New York, 11 Sept. 1967.

[48] 'Risk Factors in Coronary Heart Disease', Publication of the National Heart Foundation of Australia, Canberra, 1968.

[49] Personal Communication, Department of Customs and Excise, Canberra, consumption of tobacco and cigarettes in Australia.

[50] R. Doll and A. Bradford Hill, 'Mortality in Relation to Smoking: Ten Years' Observation of British Doctors', *Brit.med.J.*, **1**, 1964, 1399-410, 1460-7.

[51] S. B. Furnass and H. M. Whyte, 'A Student Health Survey', in press.

[52] E. M. Jellinek and M. Keller, 'Rates of Alcoholism in the United States of America, 1940-1948', *Quart.J.Stud.Alc.*, **13**, 1952, 49.

[53] E. H. Derrick, 'A Survey of the Mortality Caused by Alcohol', *Med.J.Aust.*, **1**, 1967, 914-19.

[54] 'Royal Commission into the Sale, Supply, Disposal or Consumption of Liquor in the State of Victoria. (1964-1965)' Pt 1, 27 and 82. Melbourne, 1966, Govt Printer.

[55] P. Kincaid-Smith, 'Analgesic Nephropathy in Perspective', *Med.J. Aust.*, **2**, 1967, 320-1.

[56] R. A. Douglas and E. D. Johnston, 'Aspirin and Chronic Gastric Ulcer', *Med.J.Aust.*, **2**, 1961, 893-6.
[57] Rachael Carson, *Silent Spring*, Boston, 1962, Houghton Mifflin.
[58] R. Reader, 'Prevention of Coronary Heart Disease', *Med.J.Aust.*, **2**, 1967, 546-8.
[59] Lord Taylor and S. Chave, *'Mental Health and the Environment'*, London, 1964, Longmans.
[60] J. N. Morris, 'Epidemiological Aspects of Ischaemic Heart Disease', *Yale J.Biol. Med.*, **34**, 1961-2, 349-61.
[61] C. Northcote Parkinson, *Parkinson's Law*, London, 1959, John Murray.
[62] J. Cohen, 'The Scientific Revolution and Leisure', *Nature* (London), **198**, 1963, 1028-33.
[63] A. R. Martin, 'Man's Leisure and His Health', *Bull.N.Y.Acad.Med.*, **40**, 1964, 21-42.
[64] D. Gabor, *Inventing the Future*, Harmondsworth, Pelican, 1964.

Comments

BASIL S. HETZEL The significance of civilisation for human biology is largely that of environmental change. Change is occurring with increasing rapidity as a result of the impact of technology. Technology has produced urbanisation, with profound changes in the human environment to which man is inadequately adapted. We can see this clearly by comparing life in a rural town with life in a large city.

Man's reactions to change have two major components–psychological and physiological. The psychological reaction is characteristically that of grief, for example, homesickness–a familiar experience to all in moving from one environment to another. There is a looking back on the past, a clinging to it, a nostalgia, as an alternative to acceptance of the present and the future. The most intense grief reactions are of course associated with loss of a loved one; migration may of course mean as much as death itself in final separation. The increasing mobility of urban man means that he is confronted with a more rapid succession of changes which require adaptation and make man more susceptible to the common crises of life as they occur, for example, illness, marital difficulties, problems with children's upbringing, etc.

The physiological reactions are those associated with emotional states, including more rapid pulse rate, elevation of blood pressure, changes in blood clotting time, changes in blood chemicals and secretions of hormones such as adrenalin and cortisone. Changes may also occur in various mucous membranes of the body, including the stomach and the nose, as well as in the skin. Mostly these changes are transient but sometimes they can be associated with precipitation of more serious illness such as peptic ulcer, high blood pressure, or coronary heart disease in susceptible individuals.

Adjustment to change requires time and some support for the individual so that he can leave the past and face the future. This often requires discussion and a 'working through' of feelings. Such support is normally provided by friends and family; in urban society with increasing mobility this support is not so readily available. We find that counselling and befriending services are developing in all the cities of the world because of the great need for this support for people who are not able to call on friends and family. Ability to adapt rapidly to change requires support and commitment to a purpose in life which is not always possible.

Modern civilised urbanised man is in danger of losing a sense of purpose, having to a large extent departed from authoritarian religion. The significant function of the medicine man in primitive societies is a timely reminder of the importance of a father figure in adapting to the vicissitudes of human life.

R. H. C. WELLS I would like to strike a rather more optimistic note—and to indicate that some things are not as bad as they may seem on the face of it. I think that Dr Furnass has chosen, or was given, a rather difficult subject because although we know a lot about the epidemiology of infectious disease and we have also learned a lot about genetically-determined disease, the fact is that we know really very little indeed about the quantitative effects of environmental factors on non-infectious disease epidemiology. And I would like to refer to one exception here—the case of cigarettes—because I think we can look at the data in a more hopeful manner. One can compute from quite well-collected data available that there is about one *excess* death from coronary heart disease for every 70 million cigarettes smoked, and about one excess death from carcinoma of the lung from every 300 million cigarettes smoked. That is a lot of cigarettes. If we look at this another way and take into consideration the ages at which people die from these various diseases, all we can say is that they are dying earlier than they otherwise would have done. If we make a calculation based on the available data, it looks as if for every 2 million cigarettes smoked, roughly speaking, there is one year of working life lost. Now in those circumstances, are those of us who do not smoke cigarettes in a position to judge whether the pleasure that we presume a lot of people must have got from smoking those 2 million cigarettes was worth one of them losing a year of working life? I personally do not know. I think that as far as we should go perhaps is to inform everyone of the risks.

Now our lack of knowledge about the epidemiology of non-infectious diseases has not stopped us making assumptions, and unfortunately many of these assumptions are now sanctioned by time and it has come to the point where some of them have been going around for so long that they have become accepted dogma while they are in fact still unproven. Indeed some of these assumptions still remain accepted long after the evidence indicates that they are in fact wrong. To illustrate this point I want to draw attention to two non-infectious diseases where, from looking at the data on the face value, one would have thought that something dreadful had happened and that it was probably due to the environment. Now the first of these, to take a short-time change, is asthma.

In a number of countries and particularly in Australia, round about

1950, death rates from asthma in the various age groups and particularly in the older age groups, showed a dramatic upswing; and people both here and in other countries, particularly Britain and America, were concerned to explain how this could have happened. The generally accepted explanation for a while was that this must have been due to some change in treatment of asthma and it was thought that it might be due to the cortico-steroids, despite the fact that cortisone did not become generally available until a year or so after the beginning of the trend. Well, these death rates remained very high for some years and then four or five years ago they came down again, and people began to think that perhaps the doctors and the patients were becoming adapted to the new environment imposed by civilisation, and were learning to live with these new drugs without killing one another. Now it turns out that when this is properly investigated, the whole thing is simply due to a change in classification of diseases and a change in coding practices, in the first instance round about 1950, then another change a few years ago. The whole thing is artificial.

The second example that I want to use before coming on to some general comments concerns arteriosclerotic heart disease. Now everyone knows that arteriosclerotic heart disease is more common nowadays than it used to be and that people die from it more than they used to; but is this assumption in fact true? The first thing we should realise is that coronary heart disease is a pretty recent illness; it only became coded in Australia as an official cause of death in 1931. It was only put firmly on the map of normal diagnoses by average doctors round about the beginning of World War I and consequently when we look at figures collected round about the beginning of the century and collected now it is not surprising that there were not all that number of cases recorded as being due to coronary thrombosis at the beginning of the century because coronary thrombosis was practically unknown then as a diagnosis. Not that it didn't occur; the deaths were presumably put down to something else, and as most coronary thrombosis deaths occur very quickly, within a few minutes or a few hours of the incident, they were put down no doubt as sudden death from apoplexy and a hundred other things. In fact if you take the number of deaths that were recorded as simply due to heart disease early in the century and add to these a proportion of the sudden deaths and a proportion of the apoplexies which were no doubt wrongly diagnosed, and take into account that the population age structure has changed, and further take into account that diagnoses in hospitals are only about 60 per cent accurate—there are figures to support this both in Australia and in Britain—and furthermore that in Australia coronary thrombosis is over-diagnosed in deaths in hospitals by the clinicians, one then begins to wonder whether this assumption that coronary thrombosis has become all that amount more common as a cause of death and illness is really correct. Now I do not know whether it is correct or not. I suspect that there is an increase in coronary thrombosis, but I merely use this as a second example to show that the figures may not be as depressing as they appear on the face of it.

When we start to study the epidemiology of the non-infectious diseases, try to find out the cause of these diseases, and look for changes in the environment—because it is most unlikely that short-term trends are going to be due to genetic factors—we are really putting the cart before the horse, because we are often seeking to find explanations in environmental change of non-existent changes in disease. Now having said that it seems to me that we can draw very few certain conclusions historically about the impact of civilisation on the epidemiology of non-infectious disease. But now that infectious diseases have been substantially controlled as a cause of death, even if not as a cause of morbidity, the major problem next requiring solving in the strictly medical as opposed to the medical/social field would be the chronic degenerative non-infectious diseases. Believing that we are unlikely to get very many worthwhile leaders for future action from past history, it seems to me urgent that we set about collecting now the necessary data and carrying out the necessary observations in order to be able to base future recommendations for social change and for medical change upon clear-cut and reliable data. Now until we do know more and more confidently about the effects of these processes associated with civilisation and health, I for one would resist the implied suggestion that we should give up our cars and claret and cheese and chops and live like the Chimbu, because we cannot be certain whether or not they really cause the ill-effects that have been attributed to them. I feel sure that many of these possible associations Dr Furnass has mentioned between civilised cause and clinical effect, he has really mentioned provocatively, and they should provoke us; they should provoke us to further inquiry, if in the delightful words of Eryximachus he quoted, 'We are to seek knowledge of the loves and desires of the body and how to satisfy them'.

To end on a positive note, as was suggested by Professor Stanley, I feel this is an area where we are mainly guessing, and it is about time that we ensured for future generations some first-rate and well-organised data to guess upon.

Discussion

SHATIN At the meeting of the Australian Society of Geographical Pathology dealing with the 'Pathology of Papua and New Guinea', Magarey, Kariks, and Arnold[1] presented a paper in which aortic atherosclerosis in the Territories was compared with that of Sydney. These investigators made planimetric measurements of various types of lesions which occurred in these geographic regions among carefully matched samples of both populations. They found that the differences between sexes in all regions were negligible but a comparison in the extent of lesions in the two geographic areas revealed a statistical difference at 5 per cent level in fatty streaking in the third and fourth decades, and in both fibrous plaques and complicated lesions in the fifth decade and later years. In all instances the Sydney aortas had more extensive lesions than those from the Territories. Professor Magarey felt disinclined to suggest a cause for these findings. The widely current interpretation of this higher incidence and severity of atherosclerosis in the economically developed countries than in those which are still largely undeveloped, is the high consumption of animal (saturated) fats but an even closer correlation between the dietary pattern and atherosclerosis has been demonstrated by Yudkin.[2] He pointed out that during the past century, the *per capita* annual consumption of sugar in the UK has risen, if I recollect rightly, from 7 to 125 lb. It seems to me that irrespective of which dietary items may be causative, the biologically rapid and abrupt changes in the dietary habits brought about by the neolithic and industrial revolutions will undoubtedly be shown important in the aetiology of some civilisation diseases.

1 F. R. Magarey, J. Kariks, and L. Arnold, 'Aortic Atherosclerosis in New Guinea and Papua Compared with Sydney', Abstracts of Papers presented at the Annual General Meeting, The College of Pathologists of Australia, Melbourne, 1968, 18.

2 J. Yudkin, 'Nutrition and Palatability', *Lancet*, 1, 1963, 1335-8.

LEMBERG I am grateful both to Dr Furnass and also to Professor Hetzel for their stress on what Dr Furnass has called the application of common sense and what Professor Hetzel referred to as the role of the 'modern medicine man' in helping the 'modern nomad'. Now, there is one significant difference between the nomad of old times and the modern nomad. The nomad of olden times, of the palaeolithic age, had his group to fall back on, as well as his medicine man. Now, where are the parallels in our modern society? The work of religious schools and of enlightened religious leaders could be of very great importance, and this seems to me very important when we come to deal with the imponderables which are so difficult to get access to by statistics. These imponderables are perhaps more important than anything that we can statistically put our hands on; certainly they are no less important. We can only deal with these, not by the strict scientific method, but with understanding and perhaps with a certain degree of modern psychiatry and, on a small scale, through organisations such as 'Lifeline'. But help should be given in a much greater way by establishing real neighbourhoods again—because the significant difference between palaeolithic man and the modern nomad is that the modern nomad often has no neighbours.

FURNASS I would like to say that I agree with several points raised by the speakers. First of all I think I have really been talking about industrialisation and disease rather than civilisation and disease. From the point of view of mental health, civilisation should afford opportunities for interpersonal communication which were perhaps better developed in medieval cities than in modern technological societies.

For the modern nomad, the equivalent of the Aboriginal medicine man surely is, or should be, the general practitioner, who can listen to people's problems. It may not be necessary to diagnose every disease that we see—and there may be some situations in which simple communication and understanding are just as important as a scientific diagnosis.

DUBOS Perhaps my question should be addressed to you, Professor Borrie, rather than to Dr Furnass. It concerns population explosion. There is no doubt that the introduction of DDT resulted in a spectacular increase of the population in Ceylon. But this was not due to an increase in *density* of population. What happened was that large parts of Ceylon which could not have been occupied when they were heavily infested with mosquitoes then became habitable. So that in reality there was an occupation of new country rather than an increase in population density. Similarly, the population of the United States has greatly increased in the past fifty years, but in parts of the country which were not available to man in the past and which became available to him through irrigation and all sorts of other resources. In a large measure, this is what has happened in many parts of the world. What has happened is not necessarily an increase in density of population but the occupation of new land by man. In several places, indeed, the density of population has decreased rather than increased. Most European cities, large and small, are much more densely populated than the American cities. New York is an empty city compared with Paris, Florence, or almost any other compact city of continental Europe. The problem of increasing population should always be stated in terms of occupation of new territory.

MCBRIDE I would like to make a comment on Professor Dubos's statement. He may take some comfort from the fact that while population numbers are going up, we are expanding into more land and not only increasing density. But of course there is a cost here. It is pretty clear that in about one or two hundred years we human beings are going to be about the only inhabitants of this planet, except for those animals and plants that directly concern us, and each migration into new areas is going to destroy more species of animals and plants.

NESTEL I was surprised Dr Furnass has allowed Dr Wells to attack his entire argument without any protest. While agreeing that much more investigation and more data are required before we can accept completely all the statements that Dr Furnass has made, I think Dr Wells has also been guilty of misstatement. One just wonders whether he has a shred of evidence for his statement that coronary heart disease is being grossly over-diagnosed in hospitals. I think Dr Furnass was very wise to begin his discussion in 1950 and his comparisons are between 1950 and 1966/67. By 1950 an awareness of the increase in coronary artery disease became prevalent and the facts are that in these last seventeen years there has been a very great increase in the incidence of this disorder and even in the pattern of the disease, at a time when many of the fallacies due to mis-diagnosis have largely been eliminated.

FURNASS I would like to thank Dr Nestel for bringing up some of the points I was going to raise. Despite the possibilities of statistical error which Dr Wells mentioned, most clinicians are agreed that coronary disease is increasing, particularly in the forty to fifty age group and younger. I do not think that there is any question that it is a pleasant way to die when you are seventy or seventy-five and nobody would argue that this is partly an attribute of increasing age. The real challenge is to prevent the disease in the working age group.

In terms of the possible prevention of so-called diseases of civilisation, I still think it is possible to eat, drink, and be merry and yet avoid some of the pitfalls in personal habits which beset us. The key word may be 'moderation'. Few people enjoy being preached at about diet or behaviour, although some associations between personal habits and disease seem to be emerging, for example between cigarette smoking, cancer, and coronary disease. While these may not necessarily represent a cause and effect relationship, it would seem to be a reasonable principle of health education for the medical profession to discuss the likely risk factors in disease and leave people to decide for themselves. If they are told that seven million cigarettes have to be smoked to cause one death—this will encourage them to smoke. Perhaps it is better to take the other line, if they are really convinced that smoking is potentially harmful, and point out that pipes and cigars seem to be much less harmful than cigarettes, rather than to produce an argument which suggests that there is really very little effect at all from the smoking habit.

Preventive measures which aim to improve the quality rather than merely the quantity of life may prove to be the most acceptable and effective in the long run.

BORRIE (*Chairman*) It remains for me to thank Dr Furnass and the discussants today. Professor Dubos ended on a hopeful note and perhaps as a demographer I can likewise end on a hopeful note. This may sound rather unusual because I think that demography at the moment seems to have inherited the position of the economists in the time of Malthus and Adam Smith of being a 'dismal science' in the sense that there are often dreadful prognostications of the population explosion. Like Professor Dubos, I do not like this word, but I think that one might point out that the way a demographer looks at the world is essentially that industrialised societies, not non-industrialised societies, are in a unique position; that is, unique in the sense that nowhere over human history have masses of populations ever before reached the position where their expectation of life at birth is around seventy years and where 95 per cent or more of the females born live to the end of the reproductive period.

Thus, reproductively speaking, we are almost immortal, and the addition of a few more lives will not mean anything in terms of reproductive increase—the more so as reproduction now is a matter that entails a very short span of life—on average long enough to raise just over two children, which is enough to give you the sort of growth figures which Professor Dubos mentioned to us today in the Australian environment. Demographers should emphasise more the point that this is a really unique element of contemporary industrialised human societies, and the control of death, particularly through the control of infectious diseases, is another unique aspect.

I think Dr Furnass drew a very nice trilogy by his illustration of the three major aspects of man: the nuclear explosion, the population explosion, and the 'Homo ludens'. This summarises the human situation very well, I think. The point is that the degree of control which mankind is now asked to exercise, in order to keep his reproduction rate to a very low level, is something that he has never been asked to do before and this has all sorts of implications in the field of social medicine, psychiatry and so on. If industrial man ever lets the lid off the reproductive pattern with the life span that he now enjoys, the explosion that would follow would be far greater than anything that we see around us today in the developing world. This is a totally new type of life to which he now has to adjust himself.

Dr Furnass did refer to (and I felt by implication deplored) the multiplication of scientific journals, but thinking about it, if we match population increase with a commensurate increase in literacy in the developing world, the rate of doubling scientific journals every fifteen years will just keep the reading matter up to the literate population growth! So perhaps it is not to be deplored after all. Professor Hetzel referred particularly to organisations such as 'Lifeline' and the telephone calls; I think this does illustrate in a sense the adaptabilities that can follow rapidly from technological advance, because this merely adds another aspect to the whole spectrum of the manner in which the ills of society can be coped with through technological advance. I think we have to be careful to differentiate between what is technologically possible now and what was certainly not technologically possible a few years ago in making comparisons. I will not enter into the discussion between Dr Wells and others on the question of accuracy or otherwise of measurement. I think there is a balance to be maintained here, but again as a demographer (and I am afraid we are all quantifiers in my profession) I did welcome very greatly the emphasis on the need for a great deal more care, and I am sure all here would welcome it, in terms of accurate measurements of the incidence of non-infectious disease, before we can really draw safe conclusions about mortality trends specific to those diseases. This is really an aspect that is still in a very rudimentary phase and in the next twenty or thirty years it is going to be important to get the statistical answers to some of the things that are puzzling us in the matters discussed today.

Well, it just remains for me to thank Dr Furnass for, I am sure you will all agree, a magnificent survey of the field and the very broad canvas that he presented to us, and to thank the discussants, and particularly Professor Hetzel and Dr Wells, for their remarks.

F. W. Clements

The Impact of Civilisation on the Biology of Man

Some Effects of Different Diets

There have been few significant changes in the dietary patterns of human beings since primitive man first wandered out from the primaeval forests into the adjoining savannah until comparatively recent times. Furthermore, these major changes have occurred so recently that the full effects may not be apparent for a long time.

I propose to review briefly the main dietary patterns and their possible effects on biological processes and then to consider in detail the dietetic and nutritional features of the mixed diet which is characteristic of the European-derived societies today and the effects of these diets on human biological processes.

However, before proceeding to this task, I want to make a few comments on the élite, specific groups which can be found in every society.

THE ELITE

In every community some people enjoy a privileged position. Their position, inherited or acquired, may accord them power or status and so provide them with an assured supply of food and other material goods and perhaps a labour force. The foods available to them are often the most delectable, most highly prized because of supposed inherent qualities in imparting strength, fertility, long life, etc. Coincidently, most of these foods also have high nutritional value. Some members of this élite became gluttons and consumed prodigious quantities of food, for example Charles V of Spain and Henry VIII of England.

In families there may be one or two members whose position entitles them to special foods, often of high nutritional value. Until recent times in Western societies this privilege has been reserved for the recognised 'head' of the family.

During this century, infants and children in many parts of the world have moved into a privileged position dietetically. In countries such as Australia, they are usually provided with a more adequate and balanced diet than that consumed by the members of most other age groups.

The main point I want to make here is that in every civilised society for which a reasonable degree of documentation exists, there have been, and are, people who, by virtue of their membership of an élite, have probably been better fed than the remainder of the population.

Depending on the type of diet and the margins between the élite and the remainder, the superior diet could have long-term biological effects.

DIETARY PATTERNS

The dietary patterns of human beings from the earliest times can be divided into three categories:

i mainly flesh foods
ii mainly vegetable foods
iii mixture of flesh and vegetable foods.

Mainly flesh foods

Peking man depended largely on meat[1] for it seems that about two-thirds of the animal remains which have been found in the caves, along with the skeletons of man, belonged to two species of deer, horse, camel, buffalo, and rhinoceros. The discovery of human long bones, split in the same manner as those of animals to extract the marrow, suggests that Peking man was a cannibal. Seeds of berries have also been uncovered, so it is possible that he may have consumed some vegetable foods. Neanderthal and Cro-Magnon men were also hunters and gatherers, mainly the former. Radio carbon dating indicates that a proto-neolithic people, the Natufian hunters and fishers, camped near Jericho about 7800 B.C., lived mainly by hunting the gazelle but they also gathered wild grain. Later deposits from the same sites about 6000 B.C. suggest some of the grain may have been cultivated.[2] This time approximates closely to the period of first cultivation of wheat or wheat-like cereals.

Relatively small groups of hunter-gatherers have existed in many locations into comparatively recent times. These include the Semang and Sakai of Malaysia, the Bushmen of the Kalahari Desert, the Blackfoot Indians of North America,[3] and the Eskimo.[4]

Mainly vegetable foods

Agriculture was an integral part of the life of large numbers of people in western Asia and eastern and western Europe[5] by the time the Swiss lake villages had been established. Wheat and barley were widely cultivated and the discovery of pea seeds in pottery jars suggests that the Swiss lake dwellers were acquainted with this legume and had learned its cultivation and use. These people were also hunters and fishers. Thus they could have had available a mixed type of diet with the possible dominance of vegetable foods. However, the percentage of calories derived from animal foods may have been high, because of the relatively small numbers of people in any locality and the comparative abundance of wild animals. With the small but steady increase in population and the gradual encroachment on to forest lands by human settlements the people came to depend increasingly on cereals and root crops. These crops yield many times more calories per unit of land than either domesticated or wild animals.

In many parts of Asia and Africa, at the present time, up to 90 per cent of the total weight of food is provided by cereals, while in the highlands of New Guinea sweet potato forms the bulk of the diet.

A mixed diet

For the purpose of this paper I have accepted that a mixed diet is one in which from 30 per cent to 70 per cent of the food is derived from animals, birds, or fish and the remainder from vegetable foods. Diets of this kind have apparently been consumed by isolated groups of people throughout historical times. Whenever possible they seem to have been the choice of the élite of most societies, but essentially they are characteristic of highly industrialised countries today.

The outstanding example of the consumption of mixed diet is the current dietary patterns of the majority of the people of Europe, North America, Australia, New Zealand, and some of the élite of many societies in other parts of the world. However, the comparative inefficiency of animals in the conversion of vegetable crops to human food (in the United Kingdom one acre produced four million calories of potato but only 350,000 calories of beef muscle)[6] creates a considerable price differential between foods of animal and vegetable origin.

Flesh foods make a number of appeals to human beings: their comparative scarcity gives them status and prestige; when cooked they have an agreeable taste; and the high protein and fat contents ensure that they remain in the stomach for a number of hours after a meal, thus imparting a feeling of satiety. In countries where a wide variety of animal and vegetable foods are available, as the economic condition of the community or individuals rises, the consumption of flesh foods increases, until they provide from 20 per cent to 30 per cent of the calories.

NUTRITIONAL EVALUATION OF THE THREE CATEGORIES OF DIETS

Mainly flesh foods

The hunting Eskimos derive a high percentage of their total food from flesh foods. Numerous dietary surveys have recorded intakes of between 2,800 and 3,500 calories per adult male, with approximately 40 per cent of calories provided by protein, 55 per cent by fat, and the remainder by carbohydrate. Studies of the metabolism and health of the Eskimo have shown that so long as large amounts of energy are expended collecting food, the high consumption of protein and fat does not appear to have detrimental effects.[7]

One feature of a high meat diet is that infants in the post-weaning period may not have access to enough nutritious, easily masticated and digested foods. If flesh foods are stewed for a fair length of time young children can masticate and digest them and get adequate calories and most nutrients. Also, if the food is pre-masticated by the mother and then fed to the child digestion usually presents no problem. Time and perseverance is an essential factor in this method of feeding. However, if the young child is expected to consume partially roasted food from the family meals, it may not get enough to eat or it may acquire dietetic diarrhoea.

Given an adequate supply of animals or fish we can accept that people consuming this type of diet would obtain their calorie, protein, mineral, and vitamin requirements.

Vegetable foods

Cereals. These include rice and wheat in Asia, maize in America, and the various millets and sorghums in Asia and Africa. With the exception of rice these cereals contain from 9 per cent to 12 per cent protein (rice from 6 per cent to 8 per cent) and with the exception of oats all contain 1 per cent to 2 per cent fat (oats about 9 per cent). Thus for wheat, oats, millet, and maize from 9 per cent to 13 per cent of the calories are present as protein-calories; for rice 8 per cent. In the diet of the hypothetical average Australian approximately 12 per cent of the calories are provided by protein.

The contribution a cereal or a mixture of cereals makes to the total food intake varies widely. Here we will consider the extreme situation where up to 90 per cent of the total calories are supplied by cereals. When the cereal is either wheat, maize, or millet, an intake of 600 grams, containing about 2,100 calories, will provide the individual with about 60 g protein; the same quantity of rice would provide about 45 g protein and an intake of about 520 g of oats (containing approximately 2,100 calories) would provide 62 g of protein and 45 g fat. These protein intakes are within the range of total protein found in many experiments to be adequate to maintain nitrogen balance in normal adult males.

However, cereals as a group are low in lysine, one of the essential amino acids; but where a cereal has become established as the staple, one or more legumes has been adopted as the main supplementary food item. In China it is soya, in India and other parts of South East Asia it is a number of varieties of pea, in Africa and in Central and South America it is one or more varieties of bean. In Europe from the Middle Ages into the nineteenth century when cereals—wheat, barley, and rye—provided a large part of the total food consumed 'pease' was the main supplementary food.

Apart from containing from 22 per cent to 37 per cent protein, these legumes are comparatively rich in lysine, thus complementing the cereal deficiency. Whole grain cereals and legumes are also both satisfactory sources of thiamine. With the exception of Indian millet (*Pannicum miliare*) and maize, all cereals contain adequate amounts of either nicotinic acid or tryptophane (the later can be converted into nicotinic acid in the body) to supply the co-enzymes needed in metabolism.

At this point we can conclude that when the calorie needs of adult males and non-pregnant, non-lactating females are fully met by a combination of three or four parts of cereals and one part of legume, the protein and most of the essential vitamin requirements will be provided.

Root crops. In this category are potato, sweet potato, taro, yam, cassava or manioc. Two important characteristics of these foodstuffs are the high water content, ranging from 60 per cent to 80 per cent and a fibre content of between 1 per cent and 2 per cent. The two main constituents are carbohydrates, ranging from 20 per cent to 30 per cent, and protein which varies widely as the following list of 'preferred' values show:[8]

	g/100 g
Cassava	0·7
Yam	2·0
Taro	2·0
Sweet potato	1·5 (range 0·4 to 3)
Potato	2·0

Viewed on this basis, (g per 100 g) these foods appear poor sources of protein. However, when the protein is calculated per 100 calories a different picture emerges. It must be appreciated that a staple foodstuff is, by definition, consumed in quantities which provide a substantial portion of the calorie requirements. If 2,000 calories are provided by roots, the following quantities of protein will be obtained:

	Protein g/2000 calories	Protein-calories per cent
Cassava	14	2·8
Yam	40	8·0
Taro	35	7·0
Sweet potato	34	7·0
Potato	54	10·8

It will be noted that from approximately 3 per cent to 10 per cent of the calories are derived from protein, the remainder are supplied by carbohydrates. Cassava is clearly inferior to the other crops and will be omitted from these discussions.

Analyses of the various roots have shown that the amino acid content and the ratio of each essential amino acid to the total essential amino acids of the proteins are closer to the Reference Protein than the proteins of cereals.[9] In other words there is not a marked deficiency of any of the essential amino acids as with cereals. So far as I know legumes or any other protein-rich food are not consistently eaten when roots are the staple food.

It is noteworthy that where either cereals or root vegetables constitute the staple foodstuff, appreciable quantities of green leaves are usually consumed; they contribute small amounts of protein but considerable quantities of calcium, iron, and carotene. Provided the individual can consume the necessary bulk of root vegetables to provide between 2,000 and 2,500 calories, together with an appreciable amount of green leaves, the protein, calcium, iron, retinol (vitamin A activity), thiamine, and vitamin C requirements will usually be obtained.

The large bulk, the high water content, and the fibre in these foods may be conducive to prolonged digestion with a prolonged low level absorption of nutrients from the intestines. If this is so, meals of these foods would approach more closely 'nibbling' in contrast to three large meals of a highly concentrated mixed diet eaten at regular intervals through the day. Frequent small meals have been shown to be associated with more efficient utilisation of the proximate constituents, especially protein.[10]

Clinical appraisals suggest adults who obtain an adequate calorie intake from cereals plus legumes plus green leaves, or root vegetables plus green leaves, appear to be adequately nourished.

For infants and children the story is different. In societies where cereals or root vegetables are staple foodstuffs there is usually no food which is entirely satisfactory for use in the weaning period and for young children. The traditional practice is for infants to be breast fed well into the second or even third year. In many communities where polygamy was, and still is, practised, a wife was not sexually available to her husband while she was lactating. This taboo ensured that many infants would have part of their calorie and nutrient requirements provided by breast milk for a year or more. Supplementary feeding with one or more of the family foods was often delayed until the child was one year old or until it had cut some teeth.[11] The food offered would be selected from the family meal. This may be prepared in a special way, premasticated by the mother or fed directly from the family pot.

Although milk contains about 88 per cent water, the formation of curds in the stomach allows an early escape of much of the whey from the stomach to the intestines. This enables the infant to consume much more milk than would normally 'fill' the stomach. Although an infant may drink 220 cc during a meal, the stomach usually only contains a percentage of this amount by the time the infant has finished drinking; it may be less than half the amount consumed.

Cooked cereals and cooked roots contain from 70 per cent to 90 per cent water. Much of this water is in physical and chemical combination with the carbohydrate, protein, and fibre and will not be released until the digestion of the carbohydrate and protein has been at least partially completed. A significant part of the meal will not pass into the intestine from the stomach in the early stages of digestion. The nutritive value of a cereal or root crop meal would be significantly lower than a milk meal of the same volume.

Mixed diets

The modern mixed diet has undergone a number of significant changes in the last 350 years; many of these have been highly desirable nutritionally. From the Middle Ages to the beginning of the eighteenth century the diet of all but the wealthy consisted of bread, usually made of a mixture of wheat, rye, and barley; peas; salted bacon; cheese and milk. Small quantities of vegetables were eaten and fruit only in season. The affluent ate, in addition, meat, poultry, venison and other game, often in considerable quantities. The diets of the peasants were adequate by modern standards, (except for vitamin C; scurvy was relatively common) and were nutritionally superior to those eaten by many people today.

The seventeenth century saw the beginning of deterioration in the nutritive value, as well as a number of improvements.[12]

The rapid increase in the size of towns and the appearance of new towns around sources of fuel, often on relatively infertile land, presented problems of food supply. At times grain production did not keep pace with the rise in population and adulteration of flour became common. Towns required large quantities of meat, which was brought in on the hoof. However, the meat was of poor quality and often decomposing before it was sold. Vegetables and fruit were more readily available than meat and relatively cheap so that there was a gradual increase in the amounts consumed, especially by the artisans in the towns.

The nineteenth century saw a number of substantial changes in diet. The remarkable industrial and commercial expansion, first in England and then in other parts of Europe, had been achieved at the expense of progressive deterioration in the health of the labour force. Under-nutrition or semi-starvation, which had been slowly increasing throughout the century, was complicated by the deficiency of specific nutrients. Bread supplied the bulk of the calories for large sections of the population. The introduction of roller-milling of wheat about 1880 reduced the nutritive value of the bread through the removal of significant amounts of thiamine and iron. Much of the butter sold in the eighteenth century and the first half of the nineteenth century was rancid. A series of cattle plagues in the second half of the nineteenth century sent the price soaring so that a search for substitutes began. It is not surprising that unenriched margarine began to replace butter. Treacle, which had become popular early in the century, was joined by golden syrup after 1880 as the main spreads for bread by the poorer people. Cheap jams of inferior quality were also bought in large quantities.

Edward Smith, who made a study of the diet of the Lancashire cotton operatives for the Privy Council in 1862, reported that they lived largely on bread, potato, oatmeal, bacon (often very fatty), treacle, and a very small amount of butter and tea. Meat and vegetables were eaten occasionally.[13]

The consumption of milk in any form by the majority of adults had never been large from the Middle Ages onward, and throughout the nineteenth century its use was largely restricted to infants. The high bacterial count of much of the milk sold in cities and towns of Europe contributed to the high infantile mortality rate.

Sugar consumption in the nineteenth century increased greatly in all industrialised countries. In 300 years the consumption in England rose from virtually zero to approximately 50 kg per head per year.[14] The most spectacular rises over the last 100 years have been in the economically advanced countries, but in the last two or three decades appreciable increases have occurred in many emerging countries.

A significant percentage of the sugar now consumed by many people replaces other foodstuffs (cereals and roots probably) which were consumed by their ancestors. In Australia, the United Kingdom, and the

United States today approximately 15 per cent of the total calories consumed by the average person is derived from sugar.

The culmination of the progressive deterioration of physique of people in England was recognised with call-up of men for the South African war. Rejections for military service over the whole country were nearly 40 per cent while in some areas they were as high as 60 per cent. Dental caries, heart afflictions, bone deformities, and inadequate height were the main causes. In 1883 the minimum height for recruits for the infantry had been lowered from 5ft 6in to 5ft 3in and it was necessary in 1902 to reduce the minimum height to 5ft. An Inter-departmental Committee set up by the government found that poverty, leading to defective diet, over-crowding, and poor sanitation were the causal factors.[15] At the beginning of the twentieth century, the diet of the working classes and poor in most countries where a mixed diet had been traditional for several centuries had a number of serious defects. A high percentage of the calories was contributed by refined white flour, sugar, and foods with a high sugar content (golden syrup and jam), and potatoes; in some respects this type of diet approached the vegetable diets consumed in Asian countries today, except for the high sugar consumption. Unfortified margarine and animal fats were substituted for butter, the consumption of milk and cheese was low, and relatively few green vegetables were eaten. In the first quarter of this century pellagra was common among the poor in the Southern States of the United States. Their diet consisted mainly of 'grits and gravy'—corn meal and lard. Many infants who were not breast fed were inadequately fed. A survey made in Sheffield in 1900 showed that over 60 per cent of the women in the working class districts were giving their babies wholly unsuitable foods. Many were partly underfed, and most had insufficient protein. In this respect the dietary deficiencies resembled those of many infants in tropical countries today, where a vegetable diet is consumed and protein-calorie deficiency is widespread in infants and children. A similar syndrome under the name of Malnärscharden was diagnosed in Vienna in 1907.[16]

THE CHANGES IN THIS CENTURY

Three major developments which have occurred this century have brought about considerable changes in the dietary patterns of the great majority of people in the economically advanced countries. These are:

i the improvements in the standards of living
ii the extension and advances in general education
iii the scientific discoveries.

Improvements in the standards of living

A general rise in the standard of living, which began with World War I, has enabled increasing numbers of people in all Western countries firstly, to purchase enough food to meet their energy needs and, secondly, to broaden the range of their purchases to include increasing amounts of the more delectable, prestige foods which are valuable sources of vitamins and minerals. In the United States, where reasonably accurate records are available, the consumption of vegetables increased by 25 per cent in this period.[17] Improved housing with better cooking and storage facilities enabled a greater number of women to prepare better meals.

General education

Universal education has increased the awareness of a larger number of people of the possibilities available to them. It has facilitated the dissemination of information and ideas. This has probably been as great in respect of food and feeding as in any other area of human behaviour.

The scientific discoveries

Infant feeding. Around the turn of the century two important advances were the recognition that bacteriologically contaminated milk was the main cause of gastro-enteritis—the main cause of the exceedingly high infantile mortality rates—and the development of adequate, easily digested milk mixtures for the artificial feeding of infants.

Food values. Probably the most significant change in the nutritive value of any foodstuff occurred with the introduction of roller-milling of wheat at the end of last century. The newly introduced white flour became the staple food of the major part of western European populations, replacing stone-ground, whole wheat flour. This change brought about a marked reduction in the intake of vitamins of the B group and of iron, which was not adjusted until well into this century when the improvements in economic conditions permitted the consumption of a more varied diet.

The discovery of the vitamins and their roles in metabolism, the common food sources of each vitamin, and the effects of processing and cooking on the vitamin content of food has enabled diets for the healthy citizen and the sick to be scientifically planned. We know how to prevent the nutritional deficiency diseases, even if this is not done for political reasons. (Large numbers of people in rice-eating areas of Asia still contract beri beri, even though the cause is known and a simple method of enrichment of rice is available but is not used.)

We have been able to prevent a number of diseases by the judicious addition of nutrients to foods, e.g. iodide to salt and bread, iron to a variety of foods, fluoride to water, thiamine and lysine to cereals. Interest has been shown recently in the possible effects which changes in the methods of farming or animal husbandry are likely to have on the nutritional value of the product. The meat of broiler chickens contains less thiamine and more fat than birds grazed on the open range.[18] Intensively reared beef animals contained less carotene and vitamin A than cattle reared extensively.[19] Neither of these changes was considered to be of nutritional significance.

Crawford[20] has recently reported a most interesting difference in the tissue fatty acids in domestic bovids in England and free living varieties grazing either on grass lands or semi-jungle in Uganda. In the domestic animals, the proportion of polyunsaturated to saturated fatty acids was about 1:50, whereas in the free living animal it was 1:2 or 3. There was also a greater diversity of polyunsaturated acids in the free living species.

Here may be the possible effects of long-term selective breeding to produce beasts with adequate amounts of saturated fats (marbled beef). The 'scrubbers' of the Queensland outback may have their virtues when judged on this basis. Crawford wonders whether there is any relationship between these differences in the fat content of domestic animals and the prevalence of atheroma in human beings who eat this food.

Food preservation. Improved methods of preservation, particularly refrigeration, deep freeze, and freeze drying have enabled food to be stored and transported over great distances thus greatly increasing the range and availability of many foods. The outcome has been a more varied diet for a larger number of people. In general these methods of processing and handling of foods do not significantly affect the nutritional value.[21]

Convenience foods. These are foods which have been prepared to an advanced stage towards consumption by food processors. They require little or no preparation and cooking by the housewife; e.g. deep-frozen fish fingers, TV dinners, most canned food, prepared desserts, pre-cooked cereals.

In some instances the nutritive value of the article may be reduced by the processing; however, the changes in individual items are not significant, although in the aggregate they could be. It is possible that in the long term, convenience foods may change materially the dietary patterns of significant segments of the population.

EFFECTS OF MODERN MIXED DIETS ON THE BIOLOGY OF MAN

The outstanding feature of modern times has been the gradual extension in many countries of the dietary pattern of a few, the élite in earlier societies, to the great mass of the population. This has not occurred suddenly but has taken place over the last 150 years and it is still going on. The effects are only slowly becoming apparent and there is no reason to believe that all the possible effects have been noted.

The range of mixed diets being consumed today is considerable; however, they fall into a number of categories according to adequacy in calories and essential nutrients and the percentage of calories provided by protein, the various fatty acids, and by starch and sugars.

Diets which are markedly deficient in calories and some nutrients can be detected with a high degree of confidence; diets which provide a constant excess of calories are revealed clinically in a comparatively short space of time. However, the ideal ratio of calories which should be provided by protein, fats, and carbohydrates is still a matter of considerable controversy.

The physical character of the diet, especially the firmness and amount of roughage, may also have long-term effects.

I have selected conditions at the end of the nineteenth century as the base line against which to make comparisons. A few details of life, useful for comparisons, are available for that period, but unfortunately details of health and nutritional status, as opposed to disease patterns, are meagre. The food patterns, environmental sanitation, and disease patterns in the slums of the industrial cities of England at that time were not vastly different from those prevailing in many parts of the tropics today, where mainly vegetable diets are consumed and poor sanitation prevails. For these reasons it is permissible to use results of current studies in the latter as the basis for comparisons with current conditions in populations who have for some generations consumed a modern mixed diet.

The following biological characteristics will be discussed:

growth
biochemical maturation
reproductive efficiency
the degeneration of certain tissues
tooth structure and gum health.

Growth

Over the past 100 years the mean heights and weights of children at all ages have increased each generation.[22] In addition the mean heights of adult men and women have also increased, but this increase is much less than the increase in children.

An increase in the mean height or weight of boys or girls of a specified age between two points in time could be due either to larger measurements for most children or to a reduction in each successive cohort of the numbers of children of small stature and low weight. Both changes have undoubtedly occurred. However, there is clearly an upper limit to rates of growth: an individual cannot exceed his genetic potential in height but he may put on an excessive amount of fat, thus affecting his body weight. In recent times there have been only slight increases in the mean height of children from the better socio-economic groups in America, indicating that in previous decades the majority in this class achieved their potential.[23] The gradual disappearance of children whose growth has been stunted is probably the major factor in the change in mean heights.

The change in the mean height of adults can be due to a greater percentage of people reaching their genetically determined height. A long series of measurements of Norwegian men gives a good indication of the changes in growth rate, as the following figures show:[24]

1760-1830 = 70 years — 0·4 inch or 0·057 inch per decade
1830-1875 = 45 years — 0·6 inch or 0·125 inch per decade
1875-1935 = 60 years — 1·5 inch or 0·25 inch per decade
1935-1965 = 30 years — 1·0 inch or 0·33 inch per decade.

In Europe the diet of labourers began to improve about 1815 but substantial changes did not occur until about 1850; although the improvement was rapid from that time onward, there were periods associated with depressions and wars when the rate slowed up.

The causes of these changes are environmental. The most obvious is better cellular nutrition, which means a more constant and adequate supply of nutrients at all growing points, whether structural or functional (enzymatic). This in turn depends upon the ingestion of adequate quantities of the right kind of foods and intact biological systems to ensure satisfactory digestion and absorption. Later it will be pointed out that chronic infection, particularly gastro-intestinal infection, interferes dramatically with nutritional status. Periods of acute undernutrition in previously well nourished children are accompanied by delayed growth, but when the food intake is restored there is usually a satisfactory 'catch-up' in growth rates.[25] When chronic undernutrition extends over much of the growing period there is delayed maturation with stunting in adult life. The differences in height between American born Japanese and Japanese from the same ethnic group born in Japan supports this.[26] The follow-up of young children who had severe protein-calorie malnutrition showed that three and four years later their height remained well below that of unaffected children, although it was better than that of their siblings.[27]

Recently we followed up a number of infants whose weight at 12 months of age was above the 97th percentile. Their height and weight at 3 and 4 years of age were compared with those of infants whose weight was between the 40th and 60th percentile at 12 months of age. The heavier infants were both taller and heavier at three and four years, being equivalent to children about a year older. We wondered if excessive feeding in infancy had stimulated above average secretion of insulin and growth hormone which in turn had stimulated higher growth rates of soft tissues.[28] If this is the sequence of events, we are justified in asking whether a permanent pattern has been established, laying the foundations for the obesity which is a common feature in populations consuming the modern mixed diet.[29]

Growth in length of long bones (the main component in increments on height) proceeds in two stages: the deposition of a cellular matrix, which later undergoes calcification.[30] Retardation in growth is largely due to reduced cellular growth, the possible causes of which will be discussed shortly.

Growth in muscles in conditions of undernutrition has been studied.[31] Jansen compared the skin fold measurements, height and weight of groups of well nourished and poorly nourished people in West Irian. The height and weight of the poorly nourished were less at all ages than those of the better nourished. However, those children who came from a locality where the protein intake was around 16 g per day had more subcutaneous fat than those who came from a locality where the protein intake was about three times greater. The muscular growth, as measured by the diameters of the calf and thigh, were greater in the higher protein group. An adequate protein intake ensured growth in size of muscle cells, whereas in the lower protein groups, the intake of protein was insufficient to meet all protein needs and the lower priority tissues were affected, but the food intake provided calories which were deposited as fat.

Studies on African infants suggest that in the second half of the first year and for some time in the second year skeletal musculature undergoes rapid expansion in size, providing the diet is adequate. The pattern of muscle growth at this time seems to determine the extent of muscle growth during later childhood. The ultimate muscular pattern of an individual is genetically determined, but whether this is achieved must depend upon the long-term nutritional status and if the studies of McFie and Wilbourn[32] are an indication, the significant period is early in childhood when the pattern of muscle growth is being established.

From these various studies we are justified in advancing the hypothesis that some of the increase in mean weight of girls and boys at different ages in recent times is due to additional musculature. However, we cannot ignore the fact that in any collection of children today a significant number will be frankly obese due to excess deposition of fat. Mayer has shown that obese pre-school children become obese adolescents and so obese adults.[33] The desire of parents today to achieve adequate nutrition in their infants and children is an understandable reaction to the undernutrition and starvation so prevalent in many parts of Asia, Africa, and America, of which we are continually made aware by the modern techniques of communication. Clearly some parents are carrying the efforts beyond the optimal and thus may be creating a number of potential health problems in their children.

Cheek has recently made an important contribution to our understanding of growth.[34] He has used the fact that the amount of DNA per nucleus is constant at $6 \cdot 2$ μg, to estimate the number of cells in a tissue. The protein content of a muscle can be obtained and, since the ratio of protein to cell water is constant, muscle cell size can be calculated. Using these facts he has been able to show that between birth and sixteen or seventeen years of age the number of muscle cells in boys increased about fourteenfold and in girls about nine or tenfold. For girls the increase is linear but for boys it is quadratic, with the break about ten years of age, corresponding to the adolescent growth spurt. Newly produced muscle cells will increase to their potential size provided the diet is constantly adequate and the muscles are given a regular heavy work load.

For some years it has been recognised that insulin plays an important part in the synthesis of protein tissue[35] and a number of recent studies has shown that the ingestion of food incites the secretion of insulin and growth hormones in a specific pattern.[36] It is reasonable to speculate that the response of the pancreas and pituitary is quantitative. Children who have a prolonged absorption of glucose would presumably secrete extra amounts of insulin for a longer time than those with a shorter absorption period. Insulin and insulin plus growth hormone stimulate the synthesis of tissue protein.

Prolonged glucose absorption is likely to occur when the diet contains sugar and starch in a number of forms. Digestion of disaccharides with the release of glucose would occur early in digestion followed by the conversion of the starch in cereals (readily available to immediate enzyme action) to disaccharides and glucose, followed by the slower release of starch from vegetable leaves and roots. This sequence might occur several times a day in an infant. Constant over-eating could, in this way, establish a pattern of metabolism leading to early maturation and perhaps obesity.

The growth of the brain in children suffering from severe undernutrition has stimulated considerable interest in recent times. Ever since protein-calorie deficiency has been recognised as a distinct syndrome, depression and marked apathy have been accepted as part of the syndrome.[37] Several investigators claim that the brain of these children is significantly smaller than that of non-malnourished children from the same ethnic group.[38] These findings immediately raised the question whether there was an accompanying mental retardation. One study in Mexico found low scores in all fields of behaviour. It was also found that the elaboration of inter-sensory relations, which represents a set of developmental functions showing age-specific characteristics, is delayed in children with poor somatic growth due to undernutrition.[39]

It is early days in the investigations into the relationship between nutrition, cellular growth and intellectual behaviour but the observations so far made permit us to speculate that adequate nutrition in infancy and early childhood in recent times, in addition to ensuring that children will reach their potentials in respect of height and muscle development, will also enable them to reach their potential in respect of brain size and thus be capable of full intellectual development.

It is impossible to say whether a greater percentage of the population has been given the opportunity of achieving greater intellectual development in the last quarter century but it is an interesting thought.

Biochemical maturation

The average age of the menarche, in European countries for which satisfactory data are available over an extended period, has apparently fallen from 17·5 years to about 12·5 years over the last 100 years. This change has not occurred uniformly in the same country; for example, the nomadic Lapps of Norway had much the same average age of menarche (16·5 years) from 1870 to 1930, while during the same period the average of town-living girls declined by nearly two years.[40] The difference between rural and urban dwellers has been noted by several investigators.

All writers on this subject conclude that improved nutrition has been largely responsible for the progressive decline of the average age. Recent studies in a number of countries have revealed a high degree of uniformity in the average age of the menarche in girls from widely different cultural backgrounds and variable but seemingly adequate food intakes. This would suggest that the floor has been reached in the mean age of the menarche in most countries.[41]

While the age of menarche in an individual girl is almost certainly genetically determined, the actual event will be initiated by the release in adequate quantities of the appropriate hormones. In other words a girl must reach a definite level of biochemical maturation before the menarche will occur.

Garn and Haskell[42] have shown that the thickness of the subcutaneous fat in pre-puberty (8·5 years for girls and 9·5 years for boys) correlated well with the age of the menarche and the year of completed epiphyseal union of the tibia, itself the result of hormone activity.

Children suffering from protein-calorie deficiency appear to have delayed or reduced production of a wide range of enzymes. Abnormalities in the metabolism of the aromatic amino acids similar to those found in some of the inborn errors of metabolism have been described in these children.[43] These abnormalities disappear when the protein-calorie deficiency is adequately treated. This is undoubtedly an extreme form of delayed biochemical maturation, but these events give an indication of the effects of undernutrition on some areas of biochemical maturation.

Sexual maturation is one of the few areas which demonstrate the variation in biochemical maturation in the normal population. Although this is a biological phenomenon its consequences are almost wholly sociological. Many societies are having difficulties in adjusting to the fact that the gap between biological maturity and social/economic maturity is widening as a result of the improvement in the nutritional status of children.

Reproductive efficiency

For many years nutrition workers have been concerned about the nutrient requirements of the pregnant woman and the possible effects of inadequate diets on the progress of pregnancy, labour, and the health of the infant.

Early investigators assumed that pregnancy was a state of extreme nutritional stress, believing that the pregnant woman must consume sufficient calories and nutrients for a non-pregnant woman plus the additional quantities needed for the foetus and the extra tissues associated with the pregnancy. It was commonly believed that food intakes which are adequate for the non-pregnant woman may be grossly inadequate during pregnancy.

It is now recognised that physiologically the pregnant woman is not equivalent to a non-pregnant woman plus a foetus; she is a biological entity, having distinct physiological and biochemical characteristics. The pregnant woman retains additional protein,[44] has increased absorption of dietary iron and calcium and vitamin B_{12} and the urinary excretion of calcium, thiamine, and riboflavin is reduced.[45]

This capacity for adaption allows the non-pregnant woman who is consuming a barely adequate diet and who becomes pregnant to meet the needs of pregnancy without substantial additions to her diet. It does not, however, help a woman who has been chronically undernourished for a large part of her life to provide adequately for her foetus without considerable sacrifice of her own tissues.

We do not have any satisfactory information on the weight gains of undernourished pregnant women in Europe during the second half of the nineteenth century but we do have some information for women in a number of tropical countries today. The following figures of weight gains during pregnancy were collected by the author during two assignments with the World Health Organization:

Weight gains during pregnancy

	Taiwan % of women	Ceylon % of women
Gained 8-11 kg	25	10
5- 8 kg	25	32
3- 5 kg	30	30
less than 3 kg	15	23
Lost weight	5	5
No. of records	130	88

It would appear that a significant number of women failed to gain an amount of weight equal to the weight of the baby they produced and a few women actually lost weight. It would be unreasonable to assume that the foetus is unaffected when the mother suffers severe tissue destruction.

Antonov[46] described the consequences for pregnancy of the siege of Leningrad during World War II. Amenorrhoea was common and relatively few pregnancies were recorded, most of them in women who, for one reason or another, received priority rations. Birth-weights were 400-500 g below the average for the periods before and after the siege. A similar story is told in respect of the blockade of Rotterdam from October 1944 to May 1945. It is estimated that adults consumed about 800 calories per day.[47] About 50 per cent of the urban women had amenorrhoea and the birth rate subsequently fell to about one-third of normal. The average birth weights of infants who spent the third trimester of pregnancy during the height of the famine was about 240 g lower than the periods before and after the blockade.[48]

It would seem that gross undernutrition in the mother can affect the birth weight of the foetus.

Thomson and Hytten[49] correlated the birth weights of infants and the incidence of prematurity in 479 primiparous Aberdonian women with their estimated calorie intake during the latter part of pregnancy. The following figures show the effects of increasing calorie intake:

Calories	Mean birth weight kg	Premature %
Less than 1,800	3·09	8·5
1,800	3·19	5·7
2,200	3·21	7·7
2,600	3·21	4·8
3,000+	3·33	1·5

Comparisons of the birth weights of infants born to women in the same ethnic group but in different socio-economic classes have revealed that well-to-do women tend to have bigger babies than do poor women.[50]

It is now recognised that a significant percentage of infants born in populations consuming inadequate diets formerly considered premature are in fact full term-small weight babies, now described as dysmature. Numerous studies show that as the level of food intake rises in a community, the number of small-for-age babies and the still-birth rate both decrease.[51]

Over the last twenty years or so obstetricians in most Western countries have endeavoured to control the weight gains of pregnant women by diets as a means of preventing toxaemia of pregnancy. The type of diet usually recommended will also ensure the transfer across the placenta of adequate amounts of the essential nutrients. This may be the explanation for the change in the average growth pattern of infants born in the last few years in New South Wales compared with those born thirty years ago.[52] Currently infants grow much more rapidly in the first three months, even though the mean weight at three or four months may not be much greater than the mean weight of the infants thirty years ago. We might speculate that the modern infant has more adequate tissue reserves at birth which augment oral intakes in the early post-natal period.

The conclusion for this section is that in the last quarter century or so diets of pregnant women in many parts of the world, especially in developing countries, have improved significantly, due to a number of factors. These include better understanding of the dietary needs of pregnant women, improved economic conditions, enabling greater percentages of women to purchase more nutritious foods, the efforts of obstetricians to prevent toxaemia of pregnancy. The outcome has been a lowering of the maternal mortality rates, fewer premature and dysmature babies, fewer still-births, a lower infantile mortality rate, and probably more rapid growth in the early months of life.

The degeneration of certain tissues

One outstanding feature of Western communities is the high prevalence of ischaemic heart disease associated with atherosclerosis. Two features of the diets in most Western communities are the high consumption of saturated fats and sucrose.

In a series of cross cultural studies Keys was the first to show that in populations which derived more than 40 per cent of their calories from fats, most of which were saturated, there was a high prevalence of atherosclerosis and the serum cholesterol was high and rose progressively with age. The reverse was true of populations with a low consumption of fat.

Furthermore, he showed, within the same ethnic group, that, as the percentage of calories provided by fat rose, so did the serum cholesterol.[53] It is common knowledge that the substitution of polyunsaturated fats for a significant percentage of saturated fats will result in a lowering of the serum cholesterol levels.

A high intake of saturated fats is not of itself the cause of an increased prevalence of atherosclerosis and of high serum cholesterol levels. This has been demonstrated by the studies made by Gsell and

Mayer[54] of Swiss dairy farmers who consumed up to 6,000 calories a day, about 50 per cent of which was contributed by saturated fats. The amount of atherosclerosis was minimal and death rates from ischaemic heart disease extremely low. These people work hard throughout their lives carrying heavy loads up and down the steep mountains; in doing this they apparently utilise the calories as they are consumed.

The high sugar content of modern diets has been the subject of a number of papers by Yudkin.[55] He points to the close correlation between the progressive rise in the consumption of sugar in a number of countries and the increase in the prevalence of atherosclerosis and of ischaemic heart disease. It has been reported that an increase in the proportion of carbohydrates supplied as sucrose is followed by a marked rise in serum-triglycerides and a lesser increase in cholesterol concentration. The substitution of starch for sucrose in a high carbohydrate diet reduces the serum-triglyceride concentration.[56]

The relationship of high serum cholesterol or high serum-triglyceride levels, either separately or in combination, to ischaemic heart disease is a matter of much investigation at the present time and to digress further into this controversial topic would be too time-occupying. The point to be made here is that modern mixed diets with their high content of saturated fats and of sugar appear to bring about changes in the metabolism of glucose and fatty acids which in some way lead to an increase in atherosclerosis. However, other factors, such as insufficient physical activity, itself a characteristic feature of a high percentage of people consuming a Western-type modern mixed diet, appear to make a contribution.

Dental caries, periodontal diseases, and malocclusion

Dental caries. Present knowledge suggests that dental caries is a multicausal disease depending on genetic, dietary, and microbiological factors.

Dietary factors include both the chemical composition and character of the food. Epidemiological data suggest that unrefined cereals are less cariogenic than highly refined cereals. In 1941 Osborn[57] postulated the existence of a 'protective agent' in crude cereals and also suggested that sugar cane and sugar cane juice contained a substance missing in highly refined sugar.

The occlusal surface of molars and bicuspids, the promixal surfaces and the gingival third of the labial and buccal surfaces are the 'sites of election' of caries. They are also sites where debris can collect most easily and facilitate the formation of mucinous plaques. These are composed of proteins and a mass of interwoven organisms.

The effectiveness of the plaque as a barrier between solutions in the oral cavity and the surface of the tooth depends on the concentration of oral solutions. In the resting phase the saliva is free of sugar, but after a meal rich in sugar the concentration of sugar in the plaque rises steeply and the pH in a dental cavity under the plaque can drop from 7·0 to below 5·0 in a few minutes and remain down for upwards of two hours.[58]

The Vipeholm study in Sweden[59] confirmed that the prevalence of caries increased as the sugar consumption rose; the risk is greater if the sugar is consumed in a form which has a strong tendency to remain on the surface of the teeth, that is toffee and other sticky confectionery. The risk is greatest if the sugar-rich food is consumed between meals.

Modern merchandising, with the display of chocolates and confectionery in prominent places in supermarkets (e.g. beside the cash registers), and the marketing of 'TV' confectionery, presumably to be consumed while watching television, together with the increasing consumption of frozen desserts, ice-cream, flummeries, etc., which contain much more sugar than desserts set at room temperature, are designed to ensure a progressive increase in the consumption of sugar.

Periodontal diseases. Periodontal diseases are major causes for the loss of teeth throughout the world today. It is clear from fragmentary documentary evidence and paleopathological studies that man has been subject to periodontal disease since prehistoric times.[60]

Periodontal diseases have been associated with many etiological factors, including malocclusion, micro-organisms, malnutrition, specific nutrient deficiencies, poor oral hygiene, calculus. The changes, characteristic of the disease, appear to be correlated with aging, which has prompted the suggestion that age is an important factor in the etiology.

Periodontal Index scores for persons aged forty to forty-nine years in a number of populations throughout the world have been studied by Russell.[61] He showed a wide range of prevalence, the lowest being in white Americans and the highest among primitive groups in Vietnam and similar localities. Russell concluded that periodontal disease is not a manifestation of a specific nutritional deficiency, but he could not rule out the possibility that a combination of dietary factors may be causally associated with the disease.[62] Despite much research into this disease, the full etiological picture has not been determined.

Malocclusion. The food of Stone Age man was tough, hard and probably highly gritty from the cooking methods and it contained much indigestible bulk. He spent a much longer time chewing his food than modern man. We have already pointed out that one of the characteristics of modern diets is the softness of most foods, even flesh foods are usually prepared and cooked in such ways as to render chewing unnecessary.

The prolonged chewing and the nature of the food eaten by primitive peoples caused both the deciduous and permanent teeth to undergo extensive and rapid attrition. This occurred on three surfaces, the occlusal, proximal and incisal surfaces, and resulted in the continual reduction of the size of the teeth. As a consequence teeth changed shape and position with age. There was a progressive mesial migration of the molars, which allowed the much earlier eruption of the third molar. In primitive people with extensive attrition the third molar apparently erupts about two years after the second molar. At this age the roots have not completely formed.[63] In civilised man, because of non-attritional occlusion, the third molars are the only teeth that have completed the formation of their roots before eruption.

The observations of anthropologists that there was a relatively lower incidence of tooth irregularity and crowding in primitive man, compared with its higher incidence in civilised man, led to the almost certainly erroneous conclusion that there had been an evolutionary reduction in the size of the jaws. The modern view is that lack of adequate attrition in civilised man, because of the character of the food eaten, is a much more likely explanation.

With the prospects of still more soft and pappy foodstuffs in the future, particularly with the greater use of convenience foods, we can expect dental overcrowding to continue to be a troublesome problem.

CONCLUSIONS

A mixed diet, that is one in which a significant portion of the total food consumed is of animal origin, has been consumed by some human beings for thousands of years. Today it is the characteristic diet of the majority of the people in Western societies and is consumed by an increasing number of people in developing countries.

In Europe through the seventeenth, eighteenth, and nineteenth centuries a large percentage of the population, although nominally consuming a mixed diet, ate small amounts of animal foods. In this respect they probably differed little from large sections of the populations in developing countries today. There is good evidence that the poor in most Western countries at the end of last century ate diets deficient not only in calories, but also in protein and other essential nutrients.

A number of important changes have occurred in the dietary patterns of the populations of Western countries this century. Firstly, increasingly more people are eating adequate quantities of those foods which provide essential nutrients and the great majority of people have enough to eat. However, there are minorities who are not adequately fed.

The *per capita* consumption of flesh foods, milk, vegetables, and fruit has increased, due mainly to more people eating more of these foods. The consumption of sugar and fat has increased markedly in some countries.

The recent work of Crawford[64] suggests that selective breeding of cattle and sheep to produce prime beasts at early ages and intensive feeding with improved pastures and concentrates have also produced beasts which are markedly different from 'bush' raised wild cattle and perhaps 'bush' raised domestic cattle. These differences include both the quantity and quality of body fat and that dispersed through the muscles. These changes could be significant in a number of ways for people consuming prime quality flesh foods. The total fat intake and the nature of the fatty acids could be factors contributing to obesity and perhaps the degenerative changes in tissues.

Perhaps the most significant change has been the increased consumption of protein-rich foods by infants and children (mostly milk and meat).

A number of parallel changes have occurred which almost certainly bear a cause-and-effect relationship to the changes in dietary patterns. These include increased rates of growth of infants and children, earlier maturation of girls and boys, increases in the final adult height, markedly reduced age-specific death rates up to middle life, greater reproductive efficiency, but there has also been an apparent increase in the prevalence of the degenerative diseases.

The constantly high consumption during infancy and childhood of some foods, especially those which permit the rapid and constant release of glucose into the blood stream, may set patterns of both food intake and metabolic processes which persist for long periods in the life of the individual.

There is no evidence that early maturation, of itself, is beneficial to the individual. It is possible that it is sociologically undesirable in modern Western cultures. Early maturation may guarantee an adequate supply of the raw materials for antibody formation and it may provide a buffer against the stresses of daily living by ensuring that there is always a large labile reserve of nitrogen, but it may condition the individual to a particular type of diet and tissue metabolism.

Epidemiological studies in many countries suggest a causal relationship between diet and atheroma and so perhaps to coronary heart disease. Attempts have been made to incriminate specific foods but it may be that the great reduction in activity, characteristic of so many people in the countries with high mortality rates for ischaemic heart disease, in association with a constant abundance of foods, may be just as important, as Hipsley and Furnass have pointed out.[65]

Teeth, gums, and jaw structure each appear to have been affected by modern diets. The consumption of foods rich in sugar at particular times in the day has been shown to correlate closely with the caries rate; soft foods which require no chewing appear to be associated with gum diseases and inadequate development of the jaw.

If we were accountants and had the task of establishing a balance sheet of the effects of the modern mixed diet on the biology of man it would be interesting to see the magnitude of the overall credit or debit.

Summary

The first man ate venison and such animals as he was able to drive into rudimentary traps; he also probably ate berries and fruits. Many thousands of years later he learned to cultivate cereals and so established relatively permanent communities. From this time onward variable numbers of people at different times in history and in many places have consumed diets containing significant amounts of animal foods as well as vegetable foods, i.e. mixed diets.

Mixed diets have been eaten by Western-derived societies for at least the last 1,000 years, but have become characteristic of those societies in the last 350 years. The élite have always enjoyed a mixed diet, peasant farmers did during the Middle Ages in Europe, but the majority of the population seldom ate more than token amounts of animal foods after the onset of the industrial revolution until well into this century.

A mixed diet has become the characteristic diet of most peoples in Western societies and the élite in the majority of the developing countries this century.

These substantial changes in dietary patterns have been associated with greatly accelerated rates of growth in children; earlier biochemical maturation, as revealed in the earlier menarche; reduced mortality rates, especially in infancy and early childhood.

The reproductive performance of women appears to have been substantially improved, resulting in lower perinatal deaths and an increase in the average weight at birth.

On the debit side is an increase in the degenerative diseases, dental caries and malocclusion.

References

1 C. C. Furnas and S. M. Furnas, *Man, Bread and Destiny*, London, 1938, Cassell, 9.

2 G. Clark, *World Prehistory*, Cambridge, 1961, Cambridge University Press, 34.

3 C. D. Forde, *Habitat, Economy and Society*, London, 1957 (reprint), Methuen, 11, 24, 45.

4 F. Mowat, *People of the Deer*, London, 1954, Readers Union.

5 V. G. Childe, *The Dawn of European Civilization*, London, 1957, Routledge and Kegan Paul (6th ed.), 288.

6 J. R. Marrack *Food and Planning*, London, 1943, Gollancz, 224.

7 A. Hygaard, 'Studies on the Nutrition and Physiopathology of Eskimos', *Norsk. Videnskaps—Akad.Skr.Naturv Kl.*, No. 9, 1941, 176; C. G. Wilber and V. E. Levine, 'Fat Metabolism in Alaskan Eskimos', *Exp.Med.Surg.*, **8**, 1950, 422-5.

8 B. S. Platt, *Tables of Representative Values of Foods Commonly Used in Tropical Countries*, MRC Special Report Series No. 302, London, 1963, H.M.S.O.

9 World Health Organization, *Protein Requirements*, Technical Report Series No. 301, Geneva, 1965, WHO.

10 C. Cohn, D. Joseph, and M. D. Allweiss, 'Nutritional Effects of Feeding Frequency', *Amer.J.Clin.Nut.*, **11**, 1962, 356-61.

11 L. Housden, 'Infant Feeding in North-east Africa', *Brit.med.J.*, **2**, 1950, 456-7; D. B. Jelliffe *et al.*, 'The Children of the Hadza Hunters', *J.Pediat.*, **60**, 1962, 907-13.

12 J. C. Drummond and A. Wilbraham, *The Englishman's Food*, London, 1939, Jonathan Cape, 117.

13 E. Smith, *Journal of the Society of Arts*, 1864, Vol. 13, quoted by Drummond and Wilbraham (see reference 12), 394.

14 W. R. Aykroyd, *Sweet Malefactor, Sugar, Slavery and Human Society*, London, 1967, Heinemann, 10.

15 Report of the Inter-Departmental Committee on Physical Deterioration, 1904, quoted by J. C. Drummond and A. Wilbraham, *The Englishman's Food*, London, 1939, Jonathan Cape, 485.

16 A. Czerny and A. Keller, *Des Kindes Ernährung, Ernährungsstorungen u Ernährungsterapie*, Vol. 1, 2nd ed., Leipzig, 1928, Deutiche, 810.

17 United States Department of Agriculture, *Food*, The Yearbook of Agriculture 1959, Washington, 1959, U.S. Govt Printing Office, 592.

18 J. Robertson *et al.*, 'Studies on the Composition of Food, I. Some Differences in the Composition of Broiler and Free Range Chickens', *Brit.J. Nutr.*, **20**, 1966, 675-7.

[19] J. M. Harris *et al.*, 'Nutritive Value of Beef from Intensely Reared Animals', *Brit.J.Nutr.*, **22**, 1968, 21-3.

[20] M. A. Crawford, 'Fatty Acid Ratios in Free Living and Domestic Animals', *Lancet*, **1**, 1968, 1329-33.

[21] R. S. Harris and H. von Loesecke (eds.), *Nutritional Evaluation of Food Processing*, New York, 1960, John Wiley.

[22] J. M. Tanner, 'Early Maturation in Man', *Scient.Am.*, **218**, 1968, 21-7; E. M. B. Clements, 'Changes in the Mean Stature and Weight of British Children over the Past Seventy Years', *Brit.med.J.*, **2**, 1953, 897-902; H. Bakwin, 'Secular Increase in Height—Is the End in Sight?', *Lancet*, **2**, 1964, 1195-6; V. G. Vlastovsky 'The Secular Trend in the Growth and Development of Children and Young Persons in the Soviet Union', *Hum.Biol.*, **38**, 1966, 219-30.

[23] H. Bakwin, 'Secular Increase in Height—Is the End in Sight?', *Lancet*, **2**, 1964, 1195-6.

[24] V. Kiil, quoted by Tanner (see note 22).

[25] J. M. Tanner, 'The Regulation of Human Growth', *Child Developm.*, **34**, 1963, 817-47.

[26] W. W. Greulich, 'A Comparison of the Physical Growth and Development of American Born and Native Japanese Children', *Am.J.phys.Anthrop.*, **15**, 1957, 489-515.

[27] J. D. Hansen, G. L. Brinkman, and M. D. Mowie, 'Body Composition in Protein-calorie Malnutrition', *S.Afr.med.J.*, **39**, 1965, 491-5; R. F. A. Dean, 'The Effects of Malnutrition on the Growth of Young Children', in *Modern Problems in Pediatrics*, F. Falkner (ed.), Vol. **5**, 1960, White Plains, New York, 111-22.

[28] D. B. Cheek, 'Cellular Growth, Hormones, Nutrition and Time', *Pediatrics*, **41**, 1968, 30-46.

[29] S. J. Foman and G. M. Owen, 'Influence of Age, Sex and Diet on Rate of Growth and Body Composition during Early Infancy', *Proc. 6th International Congress on Nutrition, Edinburgh, 9-15 Aug. 1963*, 1964, E. & S. Livingstone, 66-74; J. Mayer, 'Obesity', *Ann.Rev.Med.*, **14**, 1963, 111-32; S. M. Garn and J. A. Haskell, 'Fat Thickness and Developmental Status in Childhood and Adolescence', *Amer.J. Dis.Child.*, **99**, 1960, 746-51.

[30] S. M. Garn, 'Malnutrition and Skeletal Development in the Pre-school Child' in *Preschool Child Malnutrition*, National Academy of Science, N.R.C. Publ. 1282, Washington, 1966, 43.

[31] A. A. J. Jansen, 'Skinfold Measurements from Early Childhood to Adulthood in Papuans from Western New Guinea', *Ann.N.Y.Acad.Sci.*, **110**, 1963, 515-31; Montgomery, 'Muscle Morphology in Protein-calorie Malnutrition', *J.clin.Path.*, **15**, 1962, 511-21.

[32] F. McFie and H. F. Welbourn, 'Effects of Malnutrition in Infancy on Development of Bone, Muscle and Fat', *J.Nutr.*, **76**, 1962, 97-105.

[33] J. Mayer, 'Obesity', *Ann.Rev.Med.*, **14**, 1963, 111-32.

[34] D. B. Cheek (ed.), *Human Growth: Body Composition, Cell Growth, Energy and Intelligence*, Philadelphia, 1968, Lea and Febiger.

[35] I. G. Wool, 'Insulin and Protein Biosynthesis', in G. Litwack and D. Kritchevsky (eds.), *Actions of Hormones and Molecular Processes*, N.Y., 1964, John Wiley, 422; C. H. Pearson 'Human Pituitary Hormones', Vol. 13, G. E. W. Wolstenholme and C. M. O'Connor (eds.), Ciba Foundation, Boston, 1960, Little, Brown, 129.

[36] K. L. Zierler and D. Rabinowitz, 'Roles of Insulin and Growth Hormone, Based on Studies of Forearm Metabolism in Man', *Medicine*, **42**, 1963, 385-402.

[37] C. D. Williams, 'A Nutritional Disease of Childhood Associated with a Maize Diet', *Arch. Dis.Childh.*, **8**, 1933, 423-33.

[38] M. B. Stoch and P. M. Smythe, 'Infant Undernutrition: Effect on Intellectual Development', *Arch.Dis.Childh.*, **38**, 1963, 546-52; R. E. Brown, 'Decreased Brain Weight in Malnutrition and its Implications', *E.Afr.med.J.*, **42**, 1965, 584-95; R. E. Brown, 'Organ Weight in Malnutrition with Special Reference to Brain Weight', *Develop.Med.Child.Neurol.*, **8**, 1966, 512-22.

[39] J. Cravieto, 'Malnutrition and Behavioral Development in the Pre-school Child' *Courrier*, **16**, 1966, 117-27.

[40] J. M. Tanner, 'Early Maturation in Man,' *Scient.Am.*, **218**, 1968, 21-7.

[41] M. E. Doster *et al.*, 'A Survey of Menstrual Function among 1668 Secondary

School Girls and 720 Women Employees of the Denver Public Schools', *Amer.J. publ.Hlth,* **51**, 1961, 1841-6; J. M. Tanner, *Growth at Adolescence,* 2nd ed., Oxford, 1962, Blackwell Scientific Publications; W. Zukowski, A. Kmietowicz-Zukowska, and S. Gruska, 'The Age at Menarche in Polish Girls', *Hum.Biol.,* **36**, 1964, 233-4; S. Madhavan, 'Age at Menarche of South Indian Girls Belonging to the States of Madras and Kerala', *Indian J.med.Res.,* **53**, 1965, 669-73; J. Towns, J. M. Johnson, and A. A. Roche, 'The Age of Menarche in Melbourne School Girls', *Aust.Paediat.J.,* **2**, 1966, 67-71; H. B. Gibson, 'Events of Puberty in 186 Tasmanian School-girls', *Aust.Paediat.J.,* **3**, 1967, 22-8.

[42] S. M. Garn and J. A. Haskell, 'Fat Thickness and Developmental Status in Childhood and Adolescence', *Amer.J.Dis.Child.,* **99**, 1960, 746-51.

[43] R. F. A. Dean and R. G. Whitehead, 'The Metabolism of Aromatic Amino Acids in Kwashiorkor', *Lancet,* **1**, 1963, 188-91.

[44] K. M. Wilson, 'Nitrogen Ketabolism During Pregnancy', *Bull. Johns Hopk. Hosp.,* **27**, 1916, 121-9.

[45] G. H. Beaton, 'Nutritional and Physiological Adaptations in Pregnancy', *Fed.Proc.,* **20**, Pt III, 1961, 196-201.

[46] A. N. Antonov, 'Children Born During the Siege of Leningrad in 1942', *J.Pediat.,* **30**, 1947, 250-9.

[47] G. C. E. Burger, J. C. Drummond, and H. R. Sandstead, *Malnutrition and Starvation in Western Netherlands,* The Hague, 1948, General State Printing Office, 20.

[48] C. A. Smith, 'Effects of Maternal Undernutrition upon Newborn Infants in Holland (1944-45)', *J.Pediat.,* **30**, 1947, 229-43.

[49] A. M. Thomson and F. E. Hytten, 'Nutritional Status, Maternal Physique and Reproductive Efficiency', *Proc. Nutr.Soc.,* **22**, 1963, 55-60.

[50] A. M. Thomson, 'Nutrition in Pregnancy and Lactation', World Health Organization working document WHO/Nut/123, March 1964 (duplicated).

[51] A. M. Thomson and F. E. Hytten, 'Nutritional Status, Maternal Physique and Reproductive Efficiency', *Proc.Nutr.Soc.,* **22**, 1963, 55-60.

[52] J. W. Bell and P. M. Lay, 'The Growth of Australian Infants During the First Year of Life: A Comparison, 1933 and 1964', *Med.J.Aust.,* **1**, 1968, 541-4.

[53] A. Keys, 'Diet and Development of Coronary Heart Disease', *J.Chron.Dis.,* **4**, 1956, 364-80.

[54] D. Gsell and J. Mayer, 'Low Blood Cholesterol Associated with High Calorie, High Saturated Fat Intakes in a Swiss Alpine Village Population', *Amer.J.Clin.Nutr.,* **10**, 1962, 471-9.

[55] J. Yudkin, 'Man's Choice of Food', *Lancet,* **1**, 1956, 645-9; J. Yudkin, 'Diet and Coronary Thrombosis, Hypothesis and Fact', *Lancet,* **2**, 1957, 155-62; J. Yudkin, 'Dietary Fat and Dietary Sugar in Relation to Ischaemic Heart-disease and Diabetes', *Lancet,* **2**, 1964, 4-5.

[56] M. J. Albrink and E. B. Man, 'Serum triglycerides in Coronary Heart Disease', *Arch.intern.Med.,* **103**, 1959, 4-8; M. Winitz, J. Graff, and D. A. Seedman, 'Effects of Dietary Carbohydrate on Serum Cholesterol Levels', *Arch.Biochem.Biophys.,* **108**, 1964, 576-9; P. T. Kuo and D. R. Bassett, 'Dietary Sugar in the Production of Hyperglyceridemia', *Ann.intern.Med.,* **62**, 1965, 1199-212.

[57] T. W. B. Osborn, 'Further Studies on the Vitro Decalcification of Teeth', *J.Dent.Res.,* **20**, 1941, 59-69.

[58] R. M. Stephan, 'Changes in Hydrogen Ion Concentration on Tooth Surfaces and in Carious Lesions', *J.Amer.dent.Ass.,* **27**, 1940, 718-23.

[59] J. A. Höjer and A. B. Maunsbach, 'Vipeholm Dental Caries Study', *Acta odont.scand.,* **11**, 1954, 195-388.

[60] I. Gluckman, *Clinical Periodontology,* 3rd ed., Philadelphia and London, 1964, W. B. Saunders, 1.

[61] A. L. Russell, 'A System of Classification and Scoring for Prevalence Surveys of Peridontal Disease', *J.dent.Res.,* **35**, 1956, 350-9.

[62] A. L. Russell, 'World Epidemiology and Oral Health', in *Environmental Variables in Oral Disease,* S. J. Kreshover and F. J. McClure (eds.), Washington, 1966, American Association for the Advancement of Science, 21.

[63] P. R. Begg, *Orthodontal Theory and Technique,* Philadelphia and London, 1965, W. B. Saunders, 11.

64 M. A. Crawford, 'Fatty Acid Ratios in Free Living and Domestic Animals', *Lancet*, 1, 1968, 1329-33.
65 E. H. Hipsley and S. B. Furnass, 'Coronary Heart Disease: Some Reflections on Its Causation and Prevention', *Med.J.Aust.*, 1, 1966, 607-21.

Comments

E. H. HIPSLEY What is impressive about the nutrition of Australians, in common with Americans, British, and some other Europeans, is the increasingly rapid rate of maturation and the progressive increase in body size with succeeding generations. We should not assume that these nutritional changes are unalloyed benefits, since there is reason to believe that some metabolic and degenerative diseases of later life, as well as social problems, may have their origin, in part at least, in these nutritional changes.

Judgments of what is nutritionally valuable and what is worthless or even harmful can only be made in terms of goals. Goals vary from person to person and from time to time. Therefore, before we can speak meaningfully of 'improved nutrition', or 'poor nutrition', we must specify the goal to which we are referring. For example, if our aim is to produce a policeman or a jockey, what is a good nutritional schedule for the one, would be a poor one for the other. On the other hand, if our aim is to produce octogenarians, both schedules could be poor ones.

Dr Clements indicated that Western civilisation has introduced profound changes in infant feeding practices. Breast feeding has greatly diminished in importance, being replaced by bottle feeding with cows' milk modified by dilution with water and the addition of sucrose. Breast milk is a low-protein, high-fat food, in which 6 to 7 per cent of the calories are derived from protein and 53 per cent from fat, whereas in modified cows' milk mixture 14 per cent, or twice this proportion of calories, are derived from protein, and a considerably lower proportion of the calories from fat. In view of the fact that the infant has a limited capacity for utilising protein,[1] and that fat has a special place in infant metabolism, you may well ask, as Dr Clements has done, whether later in life pathological states are not ultimately derived from early well-meant clinical and nutritional maltreatment.

In his paper on 'Biological Freudianism', Professor René Dubos[2] has drawn attention to the fact that the first adaptations to a situation are more important than later adaptations because they condition the latter through a mechanism of biological memory. Therefore, we must very carefully scrutinise the effects of dietary patterns in the early months of life. In fact, it is clear from much evidence, including that from children born to diabetic mothers, that we must also consider the nature of the mother's diet and metabolism.

Dr Clements pointed out that 'during this century, infants and children, in many parts of the world have moved into a privileged position dietetically'. He also points out that 'some parents are over-feeding their children'. I would like to remark that doctors and dietitians are by no means innocent in this situation, since they commonly teach that children need 'special foods', for example, protein-rich weaning foods. There is actually no scientific reason for holding this view. I believe that confusion on this point has arisen because protein needs are commonly expressed in terms relating them to body weight. It is certainly true that protein needs per unit of body weight are higher in children than in adults, for example, comparable figures recommended by WHO are 1·5 g protein per kilogram of body weight for children, and 1·0 g/kg for adults. But the fact that energy needs of children are also considerably greater per unit of weight means that exactly the same foods will meet the protein needs of both age groups equally well. Moreover, it can be argued on psychosocial and even economic grounds that it may be undesirable to encourage dietetically a child élite group.

In New Guinea, where infants are suckled by their mothers and children eat the same types of food as do their mothers, one is impressed, not by the numbers of malnourished children (although these occur), but by the numbers of healthy looking children consuming diets which on European standards would be considered grossly inadequate. It is my impression that, providing the child has the opportunity and encouragement of the mother to eat frequently and enough of the ordinary foods of the area, and providing extraneous factors such as infectious disease do not intervene, the children derive a satisfactory nutritional state. It is true that they mature more slowly, and do not grow to such a large adult size, but in many circumstances this may be no disadvantage—in fact, it could be advantageous.

This, of course, is a value judgment. In making it, I have been considerably influenced by the words of Professor René Dubos in his remarkable book *The Mirage of Health*.[3] He says:

> 'The kind of health that men desire most is not necessarily a state in which they experience physical vigor and a sense of well-being, not even one giving them a long life. It is, instead, the condition best suited to reach goals that each individual formulates for himself.'

This seems to me a very valuable concept that could lead to the introduction into biology of a useful 'theory of value'—a pressing need.

Dr Clements mentioned the possibility that the increase in rates of growth and maturation may be the outcome of abundant supplies of energy substrate and protein for prolonged periods each day. This 'time factor of energy exchanges'[4] is important. Many authors have shown that the timing of meals has a profound metabolic effect, namely that eating large infrequent meals ('meal-eating') encourages energy storage as fat, whilst small frequent meals ('nibbling') discourage it. But there is another property of food that concerns 'the time factor'. This is the ease of digestibility. Meals of sugary foods, flour, or sweet potato may all have equivalent energy values, but the times at which they yield their nutrients to the tissues of the body to take part in metabolism are quite different. Sugary foods do this rapidly whilst sweet potato takes a much longer time. Finally, there is the factor of the time relationship between energy intake as meals, and the energy expenditure as activity. The New Guinean who, before he can eat his main daily meal, must expend energy in collecting food from his garden, and who eats the not-so-easily-digested sweet potato, has a vastly different metabolic pattern from the city dweller who can obtain his readily digested cordials, sweetened desserts, biscuits, and cakes with little more effort than that occasioned by dipping into his purse.

Diseases in which nutrition plays a part may result from too much (surfeits), or too little (deficiencies). Nutritional health lies somewhere between these extremes: the exact point will depend on the nature of the dominant goals, conscious or unconscious, of the particular person. I am impressed by the fact that in our feeding relationship with our environment we oscillate between hunger on the one hand and satiety on the other. I suspect that our enjoyment of life is at its best and most piquant when our various needs are in the process of *becoming* satisfied, i.e. neither *being* satisfied, nor *being* too unsatisfied. It is evident that many of the afflictions of affluent civilised man result from a more or less chronic state of satisfaction of material needs—in other words, from a state of arrivedness. Paradoxically, it seems that satisfaction of his material needs leads to undersatisfaction of his spiritual needs. So far as the need for food is concerned, I believe it is wise for the affluent man, in most circumstances, to avoid gluttony (or oversatisfaction) in all its forms.

1 P. Hahn and O. Koldovsky, *Utilization of Nutrients during Postnatal Development*, Oxford, 1966, Pergamon Press.

2 R. Dubos, 'Biological Freudianism: Lasting Effects of Early Environmental Influences', *Pediatrics*, No. 38, 1966, 789.

3 R. Dubos, *The Mirage of Health*, New York, 1959, Harper and Bros., 233.

4 E. H. Hipsley, 'The Time Factor in Energy Exchanges', Food and Nutrition Notes and Reviews No. 21, Commonwealth Department of Health, Canberra, 1964, 100.

Discussion

HETZEL (*Chairman*) Perhaps I could take the liberty of asking the first question. I think we have touched on the question of sex this morning. The age of puberty in the female seems to be a very good indication of nutrition. Dr Clements pointed out the very striking change that has occurred over the last hundred years in Europe—as so often we are indebted to the Scandinavians for precise data. I just wanted to ask Dr Clements about the medieval situation when we are accustomed to girls being married at the age of fourteen. Is it true that the menarche did occur early, perhaps only in the élite in a rural setting as opposed to later after the industrial revolution?

CLEMENTS I think you are providing part of the answer yourself in this. First of all I do not think we have the kind of information on a wide enough scale to answer your question, but you made reference to the élite, and most of the recorded facts of this kind are recorded in respect of the élite. Peller[1] published a very interesting paper some years ago on the reproductive efficiency and the infantile mortality rate in the leading families of Europe in the sixteenth, seventeenth, and eighteenth centuries. And they had infantile mortality rates, at least judged on the records that are available, which were considerably better than the remainder of the population; these low ratios have only recently been improved. They were obviously able to produce well and they were able to look after the products of their production because they were so important in the power struggle in Europe. It would appear from Peller's work that at least the women in the élite had an early menarche. I know nothing about the peasants.

[1] S. Peller, 'Studies on Mortality since the Renaissance', *Bull.Hist.Med.*, No. 13, 1943, 427-61.

DUBOS Last year there was published a book called *The World We Have Lost* by Peter Laslett from Cambridge University. The theme of the book is a reconstruction from historical documents of the ways of life in England before the industrial revolution—the world we have lost being the world as it was before the industrial revolution. Three chapters of the book deal precisely with the question that you have raised and the answer is organised naturally about that section from *Romeo and Juliet* where Shakespeare states that the proper age of marriage is fourteen. The author of *The World We Have Lost* clearly points out, however, that Shakespeare in *Romeo and Juliet* was talking about a very special class of people. From information about the date of marriage in the different social classes, both urban and rural, in Shakespeare's time, he concludes that there was a difference of at least three to four years between the different classes in the age of first menstruation, as recorded in family bibles, and indicated by the age of marriage. It seems certain, therefore, that the social environment is of immense importance in governing the rate of physiological maturation. I have read that a peculiar phenomenon seems to be taking place presently in England, namely, that the rate of physical growth, the stature at least of young men, is now much the same in all social classes, whereas there still remains a difference in the rate of sexual maturation, at least in women. In fact, one could almost guess from history that the two rates are dissociated. People of nobility, either men or women during medieval time, during the Renaissance, even though they belonged to an affluent society, were very much smaller than we are. Just try to fit an American or an Australian teenager in the armour of a medieval knight. Most medieval knights seem to have been much less than six feet tall, even though they belonged to the social élite. And this applies to women also; from the dresses in museums, it is apparent that women were very small even in the very prosperous families of the southern United States. So there seems to be some evidence that we are dealing with at least two independent factors which govern the rate of physical development and of sexual maturation.

BENNETT I do not doubt that improved nutrition has been an important factor—perhaps the main factor—in the reduced age of puberty. But if, and I emphasise the word *if*, there is a genetical component in variation of the age of puberty then there could be a reproductive advantage associated with a lower age of puberty such that if, as a result, reproduction were to occur at an earlier age, it may be possible to have, say, seven generations in a century rather than three or four. This could be a contributing factor in the progressive reduction in the age of puberty but counterbalancing factors would need consideration.

FURNASS I should like to refer to Dr Clements's reference to the possible adaptive role of polyunsaturated fatty acids in human nutrition —and to the fact that there is more polyunsaturated fatty acid in free range cattle than there is in domesticated beasts. The same is true of human breast milk, which has three or four times more polyunsaturated fatty acid than cows' milk.[1] Breast feeding has gone out of fashion in Western society. In view of the importance of polyunsaturated acids in the growth of central nervous tissues and the rapid rate of growth of the human brain in infancy, it would seem possible that the deprivation of breast feeding, apart from its psychological effects, might also be one example of starvation in the midst of plenty. This could be a situation which could be tested experimentally in rats. It is known that deficiency of polyunsaturated fatty acids can lead to skin lesions in rats and infants[2] and it might prove a useful line of research to study the behaviour of rats who are deprived of these essential nutrients.

1 Documenta Geigy, Scientific Tables.

2 A. E. Hansen *et al.*, 'Essential Fatty Acid Deficiency in Infants'. *Essential Fatty Acids,* Fourth International Conference on Biochemical Problems of Lipids, H. M. Sinclair (ed.), London, 1958, Butterworth, 216-20.

MIMS The time of onset of menstruation may have important social implications, but from a reproductive point of view what matters is how *early* females conceive; the evidence for people, other primates,[1] and even mice[2] suggests that there is a considerable period after the first menstruation or oestrus cycle when conception is less likely. Early menstruation does not necessarily mean earlier conception.

1 C. G. Hartman, 'On the Relative Sterility of the Adolescent Organisms', *Science,* 74, 1931, 226-7.

2 F. A. E. Crew, 'Puberty and Maturity', Proceedings of the Second International Congress for Sex Research, London, 1930, 9-10.

NESTEL One of the major advances in nutrition has been the improvement in infant nutrition but I would suggest that maybe we have gone too far in that social pressures undoubtedly have led to the introduction of fat babies; the baby that wins at the show is a fat baby, the baby that is in the advertisements for food products and baby powders is a fat baby. Whereas there may be some reservations about the relative importance of saturated fats and carbohydrates in producing degenerative diseases, I think that most people would agree that obesity is undoubtedly a very major factor. And there is some very interesting information available at the present time on the relationship of obesity occurring in middle age and the fatness of the baby, as in the example of muscle cells mentioned by Dr Clements. I refer to some recent work on fat cells.[1] The multiplication of fat cells appears to take place very early in life, and thereafter the increase in fat tissue is largely due to expansion of the existing fat cells. Thus, the child who produces a large number of fat cells because of infantile overnutrition is very much more likely to become obese later in life.

[1] J. Hirsch, J. L. Knittle, and L. B. Salans, 'Cell Lipid Content and Cell Number in Obese and Nonobese Human Tissue', *J.Clin.Invest.*, **45**, 1966, 976.

SHATIN In this symposium we are concerned with the impact of civilisation on the biology of man. The transition from food gathering to food production actually led to civilisation and also introduced biologically new foods. Cereals are one example of these new foods. So is dairy produce. Surely these new foods must have provided a challenge to man's metabolism, a challenge which in some cases he was unable to meet. Gluten-induced enteropathy (coeliac disease) is a case in point. This is a disease which can only occur among wheat-rye-barley-and-oats eating peoples. I understand that the mortality figures from coeliac disease, before the early 1950s, when gluten-free diet became available for the treatment of this disease, were of the order of some 15 to 20 per cent.

I have suggested that at the inception of agriculture, when wheat was a new food, coeliac disease could have been a severe and prevalent disorder, but its incidence became infrequent, as it is today, because this error of metabolism would have the effect of a lethal mutation. However, it still occurs in one out of 3,000 to 4,000 births,[1] hence it must be a genetic disorder probably transmitted as a recessive trait.[2] If these figures of the incidence of homozygous state are applied to the Hardy-Weinberg law of gene frequencies, heterozygotes would comprise 3·2 to 3·5 per cent of the population traditionally subsisting on gluten-containing cereals. These suggestions of mine appear to be substantiated by the subsequent study of MacDonald, Dobbins, and Rubin, who found an atrophic intestinal mucosa in a significant number of asymptomatic relatives of patients with coeliac disease.[3] These investigators suggested, however, that its inheritance is based on a dominant gene with incomplete penetrance. Whatever the particular mode of inheritance may be, evidence suggests that coeliac disease has an hereditary basis.[4]

I have also suggested that a heterozygous state related to coeliac disease (irrespective of the nature of the biochemical fault leading to it), may be a predisposing factor in some disease, the epidemiology of which appears to me to correlate with the dietary habits of populations subsisting on gluten-containing staples.[5]

Domestication of animals for pastoralism is another important achievement of neolithic man. Although pastoralism by itself does not lead to civilisation, it develops new foods (dairy products) which may be alien to man's metabolism, confronting it with another challenge. Lactase-deficiency syndrome probably provides the best known example of such a challenge, and I have pointed out that the use of milk by the adult is an innovation also dating back to the neolithic revolution making this infant food available for consumption even after weaning. I suggested that if lactase-deficiency is an inborn error of metabolism, it may also be a factor tending to reduce biological fitness of populations which have only recently adopted pastoralism, making milk an item of the adult diet.

If, as it seems likely, lactase is an inducible and adaptive enzyme, its persistence after weaning could be the result of evolutionary selection. This selection process could determine the incidence of lactase-deficiency in a population, and this incidence, of course, would differ among various ethnic groups, since it would depend on the time which has elapsed from the inception of their dietary use of milk, an innovation which various cultures adopt at different historical periods.

The prediction that this historic time element may be crucial in the context of the peculiar ethnic pattern of the distribution of isolated lactase-deficiency, seems to be substantiated by several subsequent studies conducted among various ethnic groups in Australia and the USA.

Almost 100 per cent incidence of lactase-deficiency was revealed in a study of recently detribalised Australian Aboriginal children.[6]

A similar high incidence of this deficiency was demonstrated in studies of healthy adult Asians in Australia,[7] and the USA;[8] while in a study—which comprised 170 patients of several ethnic groups in that country—lactase-deficiency occurred among 77 per cent of Negroes and 67 per cent of American Indians, but only among 19 per cent of Caucasians.[9]

The finding of an even lower (6 per cent) incidence of lactase-deficiency among 100 healthy Caucasians in yet another study,[10] further supports the suggestion that heterozygosity may be a feature of this error of metabolism.[11] Or perhaps the term error of metabolism should not be applied in this context to adult Asians, Negroes, and Australian Aborigines for whom milk is a new food and as yet biologically unsuitable. On the whole, milk appears to be a suitable food for most adult Caucasians, an ethnic group who historically were first to adopt pastoralism and hence first to be subjected to evolutionary selection process resulting in almost total adaptation to this biologically new food for the adult.

1 C. Carter, W. Sheldon, and C. Walker, 'The Inheritance of Coeliac Disease', *Ann.hum.Genet.*, 23, 1959, 266-78; C. M. Anderson, personal communication, 1964.

2 R. Shatin, 'Preliminary Report of the Treatment of Rheumatoid Arthritis with High Protein, Gluten-free Diet and Supplements', *Med.J.Aust.*, 2, 1964, 169-72.

3 W. C. MacDonald, W. O. Dobbins, and C. E. Rubin, 'Studies of the Familial Nature of Coeliac Sprue using Biopsy of the Small Intestine', *New Engl.J.Med.*, 272, 1965, 448-56.

4 M. D. Milne, 'Hereditary Abnormalities of Intestinal Absorption', *Br.med.Bull.*, 23, 1967, 279-84.

5 R. Shatin, 'Multiple Sclerosis and Geography: New Interpretation of Epidemiological Observations', *Neurology* (Minneap.), 14, 1964, 338-44; R. Shatin, 'The Epidemiology of Rheumatoid Arthritis and Human Ecology', *Acta Rheum.Scand.*, 11, 1965, 161-8; R. Shatin, 'Lactase Deficiency in Uganda', *Lancet*, 2, 1966, 498;

R. Shatin, 'The Transition from Food-gathering to Food Production in Evolution and Disease', *Vitalstoffe Zivilisationskrankheiten*, **12**, 1967, 104-7.

[6] R. B. Elliott and G. M. Maxwell, 'Lactose Maldigestion in Australian Aboriginal Children, *Med.J.Aust.*, **1**, 1967, 46-9.

[7] A. E. Davis and T. Bolin, 'Lactose Intolerance in Asians', *Nature* (London), **216**, 1967, 1244-5.

[8] M. H. Chung and D. B. McGill, 'Lactase Deficiency in Orientals', *Gastroenterology*, **54**, 1968, 225-6.

[9] J. D. Welsh *et al.*, 'Isolated Lactase Deficiency: Correlation of Laboratory Studies and Clinical Data', *Arch.Intern.Med.*, **120**, 1967, 261-9.

[10] A. D. Newcomer and D. B. McGill, 'Disaccharidase Activity in the Small Intestine: Prevalance of Lactase Deficiency in 100 Healthy Subjects', *Gastroenterology*, **53**, 1967, 881-9.

[11] R. Shatin, 'Evolution and Lactase Deficiency', *Gastroenterology*, **54**, 1968, 992-3.

BOYDEN I would like to draw attention again to the fact that not everyone would accept the view that the increase in rate of growth is due to nutritional factors. I would not want to take sides on this matter myself, but if I did I think it would be on the side of nutritional factors. But it has been suggested, and perhaps Professor Bennett would like to comment on this, that genetic factors other than those that he has already mentioned in his comments, may play a role; I believe it has been suggested that the great increase in out-breeding in human populations may be a factor. In fact, I think the connection has been pointed out between the increase in stature of offspring in Europe and the introduction of the bicycle.[1]

There are three other points on which I would like Dr Clements's comments. It seems that there are two particularly important changes which civilisation has brought about with respect to diet. In the first place, there has been a fantastic increase, of course, in the number of biologically new chemical compounds which find their way into our alimentary tracts in various forms both as contaminants, such as insecticides and so on on the one hand, and also deliberately added to our food, such as artificial colouring agents, preservatives and so on. The other important change, which I think Dr Hipsley touched on, is the tremendous increase, as a result of the development of the culinary art, of technological advances and of commercial interests, in the palatability of foodstuffs, and the effect of this change on food consumption and health in general. And lastly, I would like to refer again to this question of obesity. Obesity does not seem to occur in nature, that is, in animals living under natural conditions, or in man living under pre-neolithic conditions—even when there is plenty of food available; yet it is clearly a feature of civilised society for humans and also for some domestic animals. I wonder if you could comment on why, in your view, this is so.

[1] J. M. Tanner, 'The Trend Towards Earlier Physical Maturation' in *Biological Aspects of Social Problems*, J. E. Meade and A. S. Parkes (eds.), New York, 1965, Plenum Press, 59, 61.

CLEMENTS There are a couple of items that I think I might comment on. This question of obesity has come up twice. Earlier today I was talking to Professor Dubos about the question of conditioning of people towards levels of food intake, and in my paper there is some reference to this. We have followed through three groups of babies; those who, at twelve months of age, were above the 97th percentile in weight, those who were below the third and those who were between the 40th and 60th, and we have looked at them three years later. The height and weight of the heavyweight babies were similar to children a year older. This suggests that there may be a genetic factor, but this phenomenon was pretty widely spread and it was almost universal. I suggest there is a sort of pressure of feeding which determines an individual's food intake and that this starts off by being mechanical, in terms of satisfaction of the stomach and stomach size, but ultimately becomes an endocrine factor. The early feeding thus determines a pattern which is continued throughout the life of the individual.

This question, whether it is 'good' for children to be big, is a sociological one and it is one of the most difficult problems I come up against in infant management—a considerable number of mothers equate bigness with health, and it is no use advising them about big babies, and the undesirable fact that their babies are overweight. It has no meaning to them, their only idea is that a big baby is a healthy baby; and while this is quite prominent amongst the new Australians, it is very much an Australian characteristic.

Intentional and non-intentional food additives is a subject all in itself, and a whole symposium could be devoted to it. Palatability is largely, but not entirely, a matter of sugar, and one of the characteristics of the modern age is the number of desserts and other foods that we eat that are prepared and set in refrigerators. The formula for these foods contains up to three times as much sugar as a similar dish prepared and eaten at room temperature. In order to stimulate the taste buds with a frozen food you have got to add this much sugar.

I do not want to get into a discussion at this hour on the matter of lactase deficiency in the Australian Aboriginal children, but we in Sydney do not agree with the Adelaide group. We suggest it could be an acquired factor due to long-standing chronic dysentery in the Aboriginal babies and children.

6 Glen McBride

The Impact of Civilisation on the Biology of Man

Social Adaptation to Crowding in Animals and Man

ANIMAL SOCIETIES

Animals evolve. They do so by converting various energy sources in their environment into their tissues for their essential activities. The energy harvested may be used for body growth and maintenance, for the production of offspring, the rearing of these young and their preparation for a successful adult reproductive life. In this way the spread of the parents' genes into successive generations is ensured. Animals may also 'bank' energy in various ways, by storing standing feed on territories, in hoards, or in fat, or by investing energy in the production of nests or burrows which enable energy to be conserved later. One use of energy in particular concerns us. This is the energy cost of regulating the distribution of available energy among the members of the species.

The way animals distribute energy within species varies. Most species organise into some form of society, involving spacing behaviour between individuals, and some form of control over the fixed or personal space around them. This behaviour makes possible a minimum definition of a society. Animals form a society when they are distributed non-randomly in physical space, and when the pattern of their distribution arises from their spacing behaviour, approach or withdrawal, to conspecifics, and not because of fluctuations in the environment.[1]

Some animals remain solitary in the organisation of their societies by attaching themselves to fixed areas, and maintaining exclusive control over this space in a territorial system; others control a personal sphere around them as they move over the area, the home range pattern. Animals of other species aggregate in groups of various types and sizes, from pairs up to many thousands. Grouping is a spatial concept, implying that members of a group are separated by smaller distances than are individuals from different groups. A group is thus distributed over only a small portion of the area it occupies. The group area may be either a territory or home range. These are the so-called gregarious species. Still other species aggregate in colonies of small nesting territories for part of their activities during the breeding season. Here there is some variability in the extent to which these species form groups away from the nesting colony.

Within their species societies, the animals face two types of problem. The first of these concerns their relationships with other species, generally food or predators, but sometimes there may be interspecific territories,[2] or perhaps symbiotic or commensal relationships. The second type of relationship is associated with the conspecifics, largely concerned with the distribution of resources of some type among them. Both solitary and gregarious animals make a primary distribution of living resources in terms of fixed space, as territories or home ranges. This means that gregarious animals have two types of relationship with conspecifics, those with group members, and those with members of other groups. There is generally high aggressiveness between groups, concerned with the maintenance of spacing and exclusiveness of group memberships. The second set of relationships, those between animals within a group, are of particular interest in the present context, because they constitute a basic set of social adaptations to crowding, or to life within an organised aggregate.

Life within groups takes a multitude of forms, but all seem to be variations on a limited number of themes.[3] The general patterns are as follows: (a) Intraspecific aggressiveness is regulated into a dominance hierarchy, with submission which does not necessarily involve flight. Overt aggressive or agonistic behaviour is rare. The group is, in fact, a unit of lowered and regulated aggressiveness. Most dominance behaviour is seen as priority in the use of personal space around individuals. Only when some resource is limiting, particularly space itself, is agonistic behaviour used to apply spacing priorities around the limiting resource, or more simply, to drive the subordinate neighbour away. (b) There is a co-ordination of activities within the group in a series of social subphases, associated with such activities as movement, resting, alarm, feeding, or body care. The group takes up a different spatial arrangement during each subphase. There is regulation of behaviour of each animal during the subphase, so that the repertoire of interactions is limited; for example, one seldom sees agonistic or sexual behaviour in alarm or movement subphases. The term role is used to describe the repertoire of behaviour of each animal in each subphase. It covers both the range of interactions and the spacing responses which determine group structure. (c) The normal limits of movements of individuals are between the personal distances of neighbours and the social distance, that is, the maximum distance any animal will move from the group. This is called the living space. One form of the personal distance is the individual distance,[4] which is the personal distance during the resting subphase, when it is a constant. Neither the personal nor social distance is fixed, but each varies with the subphase, and thus with the roles of each sex and age class of animals in the group. Within personal distances one sees interactions, and these include dominance and submission, in fact these are *only* seen within the personal distance. Normally subordinates submit by keeping clear of the personal area. Beyond the social distance animals become distressed and show lost behaviour. (d) Group solidarity tends to be initiated and maintained by a range of behaviour, collectively called affiliative or integrative (originally called socialising behaviour).[5] These behaviours may involve mutual reinforcement, as in allogrooming, allofeeding, and play, or may lower disruptive forces between animals as in the formalisation of agonistic behaviour, or the avoidance of strangers within large groups.[6] Animals in groups have adapted to gregarious living and are not considered to be crowded.

CROWDING IN ANIMALS

Crowding, as we understand it, may occur in all kinds of animals, and may have a different significance in each type of society. Crowding normally arises from an increase in numbers, and thus of density. It may also occur by a concentration of animals in a restricted area, as around water in a drought, or it may occur when a group becomes abnormally large.

It should be clearly recognised that animals occupy two kinds of space,[7] that which they displace physically, and an area of social space around them. It is this second type which interests us in the study of crowding, for animals are seldom crowded so that they press physically against each other, as humans occasionally do in some public transport systems. There appear to be three main categories of social space, the fixed territory, the personal sphere extending round solitary animals (or groups) on home ranges, and the personal field of gregarious animals (earlier called social force field). The personal sphere is an area extending all round an animal while the personal field appears to be an area around the face, extending further in front than to the sides. The existence of these personal areas (sphere or field) can only be detected by the responses of animals to others approaching to the personal distance (the limits of the personal areas) in any direction.

Animals do not normally enter the personal areas of others. When two animals approach closer than the personal distance, there is an alerting response, and an interaction begins. Most commonly there is simply a mild threat by the dominant of the two and an avoidance by the subordinate. Dominance is used to keep personal areas free of subordinate neighbours[8] while avoidance or flight to the social distance (or the territorial limit) is the most common form of submission in animals. Naturally there are many types of interaction which are not agonistic, but there is always the alerting response and usually some sort of conciliatory behaviour associated with further approach; courtship is an example.

The necessity for these two types of space was well illustrated in a flock of seven turkey males in a pen which was progressively reduced in area to approximately one square metre. The birds then stood around the fence facing outwards; all thus kept their personal fields free of neighbours, an intriguing solution to the crowding problem!

When animals are crowded, they find difficulty in avoiding personal areas. It appears from observation that the personal areas are reduced as the available area is decreased, for the animals become habituated to the presence of neighbours at close distances. Nevertheless crowded animals are constantly alerted to neighbours, and the frequency of agonistic interactions increases. Most of the observed movements are concerned with spacing and its maintenance by agonistic behaviour. Perhaps this is why the dominancy hierarchy attracts most attention in small crowded groups, and indeed has often been considered an artifact associated with close restraint.

It should be emphasised that animals are not necessarily harmed physically by each other in crowded conditions. Nevertheless it appears that the constant alerting and mild agonistic behaviour can become stressful, inducing strain or tension in crowded animals.[9] It is seldom that any single stimulus is stressful in animals, but stimuli can become stressful if repeated constantly. The stressful effect of crowding only becomes meaningful when we recognise the existence of this social space.

CROWDING IN ANIMAL SOCIETIES

Experimental studies have shown that the crowding of solitary animals on territories leads first to a breakdown of the territories into overlapping home ranges, and later animals form an aggregate with a dominance hierarchy.[10] In such species, problems may then occur because the normal form of submission is flight to a territorial boundary; this is not possible when such animals are crowded. Submission is behaviour which prevents or halts aggression, and without it, aggressive interactions can lead to injury or even death.

The situation in nesting colonies may be similar, because these are still aggregations of territories. Crowding means that spacing behaviour in the form of agonistic interactions takes up an increasing proportion of the activities of individuals. The result is almost continuous disturbance, desertion of nests and young, and a high incidence of reproductive failure.[11]

Crowding has various effects in gregarious species. Because the concept of group is spatial, we are describing numbers of separated aggregates of affiliated animals. Accordingly crowding can refer either to increased contacts between groups on territories or home ranges, or alternatively to crowding within groups. Where the former is indicated, one would expect an increase in intergroup contact, and a breakdown of intergroup spacing mechanisms. Perhaps the effect would parallel that of crowding of solitary individuals, or of nesting pairs (groups of two) in colonies. If this were so, the final result would be an intergroup dominance hierarchy of the type described by Southwick[12] in rhesus monkeys. Crowding of groups could occur over the whole range occupied by the species, but one would expect the major effects to be seen when groups aggregate in restricted areas, around water or concentrations of feed.

Crowding within groups is a special type of situation, for groups, by definition, occupy only a small proportion of the space available to them. One common situation is the use of artificial restraint by man in his domestic animals. Another form of crowding can occur in the wild when the size of a group becomes abnormally large.

Group size normally increases as a result of reproduction, and in some species the numbers appear to expand almost indefinitely without serious disruption within the aggregate. Such species appear adapted to life within large groups and these adaptations are, as yet, poorly understood. The problem is solved in large flocks of chickens by home range behaviour by each hen, so that each lives in a small part of the flock among well recognised neighbours.[13] Constancy of neighbours might well be the solution used by other species living in large aggregates.

More commonly there is a characteristic range of group sizes for each species. Some systematic variability appears to depend upon environmental conditions, for example, Cloudsley Thompson[14] suggested that group size is often larger in open savannah than in forests. In species in which group numbers are limited, there is some mechanism to limit expansion. Young animals reaching sexual maturity may be extruded, or groups may divide when they reach a maximum size. Group splitting has seldom been observed in detail, though Southwick[15] described a case in rhesus monkeys and McBride[16] observed flock division in feral fowl. In this case seventeen females were attached to the dominant male, and a similar number of mature young males associated with the flock. The females lived within the personal sphere of the male, protected from the attentions of subordinate males. Subordinate males normally only entered this area (radius about 6 metres) when in a submissive posture. At dusk, the radius of the personal sphere dropped to under one metre as the young males aggregated closely around the flock. The crowded females then ceased all normal activities. If the alpha male attacked a subordinate, other subordinates would immediately chase and rape the hens. The solution occurred when the beta male joined the flock and the two males were together able to provide adequate protected space for the hens to resume normal activities. Adaptation proceeded further by the subdivision of the flock.

Another form of crowding may be induced in experimental flocks of hens by injection with testosterone. The effect appears to be due to an increase in the personal fields and a consequent increase in agonistic interactions between hens.

Since crowding is related to density, it should be considered in relation to normal fluctuations in space requirements and thus density in animal aggregates. Animals divide social labour in time. Firstly they reorganise their societies regularly in major changes or social phases in the breeding and non-breeding seasons. Generally spacing requirements are maximum in the breeding season, and animals become more aggressive to conspecifics of the same caste.

Within these major social structures, gregarious animals reorganise regularly into social subphases, each with characteristic personal and social distances for each caste of animals. It is these distances which determine the density of animals within the group, or the amount of area occupied by the group. Thus the idea of density in terms of animals per unit area is not an efficient indicator of crowding unless it is compared with the normal density of the subphase. This was illustrated by the density situation in the feral fowl flock at dusk and during the day.

The degree of co-ordination of gregarious animals varies between species. There appear to be two different types of grouping or affiliative systems; one we may call federation and the other group expansion. The federal system describes aggregates which arise from the development of affiliations between individuals or smaller groups, while group expansion describes a group which grows from a family by the retention of offspring. The subphase system appears more variable in federated aggregates, and seems more efficiently regulated within subgroups. Membership often varies as individuals or subgroups enter and leave. The subphase system is better co-ordinated in expanded groups. Co-ordination of roles, activities, and spacing allows a high density without crowding.

Another problem affecting the effective degree of crowding in an animal aggregate is its caste composition. Caste is an important subdivision of social labour for different but complementary ways of life within societies. For example, sex is a basic division of labour in reproduction, but in most species it is much more than that, since the sexes are behaviourally organised as castes. Here the concept of caste as an organically organised division of labour is borrowed from the social insects, and used in a different sense than by sociologists. Sexes are different castes in many, but not all vertebrate species, depending upon whether the division of social labour is equal or complementary. The other caste division is by age, and animals pass through a series of castes until they mature.

Of interest to us here is the fact that each caste normally has different space requirements. Thus the effective density of any aggregate depends upon the castes present and their proportions. Again, this was apparent in the feral fowl example, where the disruptive pressures developed as a result of the relatively large number of adult males present. It is probable that the process of extrusion of males from social groups as they reach sexual maturity is a social adaptation to crowding, for males generally have greater space requirements than females, and their competition is disruptive.

HUMAN SOCIETIES

The society of any species is the product of the behaviour of the individual animals. The society evolves as the behaviour evolves. Animals live in societies, and the environment to which they adapt by natural selection is largely a social one. We have already discussed some of the societal forms which have evolved in various species. It can be argued that these various patterns may all be described in terms of four main dimensions of societal evolution,[17] caste, phase, group structure, and spatial organisation. If we classify animal societies on scales on each of these dimensions, it is possible to see some general pathways of societal evolution. Using these it may be possible to reconstruct the path of human social evolution. Most of the stages in this scheme are represented in many other animal species, though the details will be avoided because of limitations of space. An understanding of the evolution of man aids comparisons with other species in such matters as crowding. The proposed evolutionary steps are as follows:

a. Pairs with young on territories. This is an extremely common and simple form of social organisation. In the primates it is best represented by the gibbons. The group size is maintained by the extrusion of young animals as they reach sexual maturity.

b. Colonial nesting-territorial feeding. Here it is suggested that the basic family territorial system was retained but modified so that the nests were aggregated in the corners of territories radiating outwards from the nests. One might expect this stage to develop in response to a recession of forests to clumps around water, and the adoption of a terrestrial life for feeding away from the nests.

c. Colonial nesting-group feeding. Here there is a breakdown of feeding territories and federation of families into a group away from the nesting areas, possibly an adaptation to large terrestrial predators. Dominance hierarchies between families replaced the feeding territories but the nesting territories were retained. The system differs little from that of the jackdaw,[18] but is already a 'village' model,* basically a human type of social structure.

d. Division of labour in food collection. This occurred along sex caste lines with the males becoming hunters with tools, adopting erect postures which limited the mobility of pregnant females to the nesting colony area. Human language evolved at this stage.[19] Young were still extruded from families as they reached sexual maturity, but fostered affiliations between families when they mated and remained within the 'village'. One might also expect that a process of group splitting and multiplication occurred at this stage, with links between neighbouring 'villages' maintained by exogamy. Systems similar to this still exist in some hunting and foodgathering cultures. Families are similarly specialised repetitious units within 'villages', which are in turn similarly specialised repetitious units within the culture. Functional specialisation is chiefly by age and sex castes, though there is generally some specialisation within castes in human groups, perhaps only in social rank and authority.

e. Further evolution. All types of human society appear to have evolved from this basic 'village' model. Human societies seem to have evolved almost every type of grouping and spacing system found in other species, as well as many new ones. Two main processes seem to be involved, the further federation of 'village' units to larger political groups, and the development of functional specialisation within castes, a feature rarely found in other animals. Territorial patterns of family 'nesting' are widespread, but the territorial family harvesting re-evolved with the development of agriculture.

Agriculture meant that families could produce more food than they required, and this food could be stored. It was then available to support individuals, families, and later functional groups within castes specialising in complementary activities, for defence, mining, religion, entertainment, education, authority, and the so-called secondary

* The term 'village' is used here in a special sense to mean a group living area or camp with family territories which are fixed, so long as the camp itself is fixed.

industries. In modern societies, all of these activities are characteristically handled by groups, though most activities within groups are still divided between sex castes, even though this now seems irrelevant and irrational.

Other gregarious animals live within one group, reorganising into social subphases for each different activity. Humans, on the other hand, are basically group makers. The family remains the basic group, but with the development of what we call civilisation, the range of specialised functional groups has become enormous. Each person holds membership in a large number of them, typically moving between groups to change activities.

Subphase roles within animal aggregates are largely determined by caste, with some variation depending on rank. A social hierarchy is seldom simply a set of dominance-subordinate relationships, but a set of roles. Rank, particularly the alpha rank, may enter roles along with leadership, initiation of activity change, group defence, sentry duties, retrieval of lost or distressed individuals, suppression of disruptive aggressive or sexual behaviour within the group, and the expression of status by postural communication, on which flockmates organise spatially.

The role structure of human societies is even more complex, and roles are strongly concerned with the functional activities of the group. Dominance-subordination relationships are present, but directed mainly towards the control of the functions of the particular group. This is really not so different from other animals, since it is only in crowded groups that the roles of the hierarchy become concerned only with the control of spacing by agonistic behaviour.

In most large human groups, the hierarchy tends to be pyramidal rather than linear. Most of the relationships between individuals are formalised into relationships between roles. The concept of role as used in group subphases is particularly appropriate to many human groups, since they exist for only one particular activity, and simply disappear when the activity ceases, for example at the end of the working day.

Family groups, on the other hand, participate in a number of activities. They thus change organisation and roles through a number of subphases. They often change location with subphase, particularly for recreation, and so travel between subphases. Man has the ability to learn a large number of roles, and to change roles with great rapidity as a group reorganises for each activity.

Because we must join many groups, emphasis is placed on affiliative and integrative skills, to increase co-operation and co-ordination, and to reduce possible disruptive competitive behaviour and consequent tensions within organisations. Within large cities, the total aggregation may include several million people, yet individuals develop affiliations with only a small proportion of their neighbours, rather like the hens in a large fowl house. But we do this by affiliating with others in the various groups we join, and move between groups in anonymous congregations.

CROWDING IN HUMAN SOCIETIES

Crowding appears to have similar effects on both humans and other animals. Basically, high density causes frequent intrusions in the personal areas of neighbours. These fields are as important in man as in other species. This can easily be verified by thrusting one's face a few centimetres in front of that of a stranger. The result is a strong alerting for both individuals, and avoidance or conflict. Interactions are seldom intrinsically stressful unless highly competitive, yet at high frequency they do induce tension or strain. In both animals and man, the ability to regulate the frequency of interactions depends upon the ability to control space, either in the form of a personal field or some fixed area. High social rank generally gives this control of the personal environment.

The definition of what constitutes crowding is as difficult in modern human societies as it is in animals. Man lives in as wide a range of societal conditions as is found in most other species. He may live in family harvesting territories or farms, in large or small nesting colonies, in close proximity to the sites on which productive employment groups gather daily. Children may live in single sex groups (in boarding schools), as do men working in remote areas.

Crowding may be of people within groups of all types, or of the groups themselves, or even of individuals as they move between groups. Subphase also affects crowding, so that crowding may be negligible when individuals are asleep. The types of crowding which appear important in human societies are:

a. Crowding within groups. Density and frequent contact between individuals can occur in any group, at least in some subphase. Close aggregation tends to be avoided in productive groups in the various industries which serve modern societies. Crowding of employees was once common, but has now been largely eliminated, partly due to pressure by employees in trade unions, but more significantly by employers in the interests of efficiency. Employees are spaced so that they do not interfere with each other's work, and avoid time-wasting social contacts and disruptive agonistic interactions. The size of subgroups is kept small in the interests of supervision and efficiency. Co-operative rather than competitive relationships are encouraged in small face-to-face groups, since competition is agonistic and tends to increase space requirements of individuals, who feel crowded together.

Educational authorities agree that school classes can also increase their efficiency if small and well spaced. Here, however, there are fewer pressure groups strong enough to gain immediate improvement, and direct economic returns are less easily demonstrated. The result is that many schools are crowded in class size, classroom density, and also in the recreational areas.

The family is the basic human group and family crowding is relatively common in most large cities. The problem does not appear serious when families are able to live in suburbs in separate houses on separated 'territories'. Here density is seldom high because individuals are generally able to separate at will, or when together, to organise in fixed ways to avoid interference. This does require a house large enough for the family, though the surrounding land may be used in suitable weather. Crowding within families is a greater problem in tenements and apartment houses, particularly when there is poverty.

b. Movement between groups. Modern man normally lives in a family, and joins an employment group daily. He may visit other groups during the course of a week. On weekends and holidays he may move the whole family for a recreational subphase at the beach. Physical distance is involved in each movement between groups and subphases, and modern transport makes this sort of life possible. Nevertheless travel is time-consuming and conditions are usually crowded, whether on public transport or in private cars. Crowded travelling conditions, in large anonymous congregations, is exhausting and can contribute to the development of tension and strains in individuals.

c. Crowding between groups. The most serious problem appears to be the crowding of families, usually in cramped tenements, often with shared facilities. Pollution control is often poor, with excess noise, light, and smell. Families in such close contact suffer constant disturbance, interference, and often conflict.

Since members of each family have many affiliations with other groups, crowded family conditions may lead to the development of more favourable relationships outside the family, with consequent diminution of bond-servicing affiliative behaviour within the family and a further deterioration of family life. When the parents break away from family life, then the children also turn outside for close affiliations, perhaps to street gangs. These are all social adaptations to crowding, but considered socially undesirable. The situation is remarkably similar to the crowding in bird nesting colonies.[20]

When the spheres of influence of productive groups overlap, then they may come into strong competition, with consequent stress on individuals composing them. Competition is agonistic in whatever form it appears. Mild formal competition is seldom stressful in either animals or man, though strong competition potentially is.

At the highest level of societal organisation, whole nations come into competition. Here we accept the term aggression for their behaviour and describe the conflict as war, hot or cold; either is stressful.

We can see that there are many forms of crowding in modern societies; all may be stressful, though there are also many forms of stress unrelated to crowding. Societies have made many adaptations to crowding. Some aim to reduce the crowding directly, and others reduce the strain it generates in individuals. We may now consider some of these.

SOCIAL ADAPTATION TO CROWDING

While man has faced some totally new forms of crowding, it is of note that there are many more situations which are common to other species. Also the new forms of crowding are matched by ingenuity in developing a wide range of adaptations to both new and old problems. We have nevertheless arrived at many of the same solutions used by other species.

Man regulates his intraspecific aggressiveness into formal channels of competition, mainly during his early period of socialisation. Dominance is modified into well-defined roles and interrole relationships. We also co-ordinate the activities of most groups, so that all work, or all rest, or all play or relax. We place emphasis on the development of our considerable affiliative skills to develop and maintain co-operation in groups of all types.

Man evolved in small nesting colonies, and now often lives in vast colonies or cities. He adapts by limiting his affiliations to members of those groups in which he must live, work, or chooses to join. He must still come into contact with many strangers and has developed roles to adopt to allow intermingling without intrusion into their personal areas or privacy. Should an intrusion occur, then he has a few formal interactions which enable both to return quickly to the anonymous state.

The major type of crowding in anonymous groups is in transport between affiliated groups in public transport. This is a peculiarly human phenomenon. We space our industrial sites in relationship to living areas and develop public transport to make possible the movement of large numbers of people between groups. In doing this, however, we often cause the aggregation of large numbers of people in cramped conditions, and this leads in turn to even greater crowding in transport. The private car has given privacy and some degree of isolation during transport, while allowing people to live more spaciously at greater distances from work. But this in turn has generated even greater problems, for it has also made it possible to move between even more groups. The roads become crowded with cars and this becomes eventually as stressful as is public transport. The solution of this problem requires enormous resources, both of space to create highways and wealth to build them. There does not yet appear to be a simple solution.

The family nest is basically a shelter for people, separated from other people in a territorial pattern. But the problem of spacing within the family is also important. Beds are not for the use of any member of the family, but are highly personal, and regarded as territorial property which is totally accepted. This is carried further by the subdivision of living space by walls to make rooms, designed for various activities or subphases of the family group, or as private territories, with some restrictions on entry. There is no *a priori* reason why we should organise our living space in this way, for it is basically quite inefficient, but it is hardly questioned. It appears to be a rather basic adaptation of our environment to our nature, which is concerned with the maintenance of personal living space. We also use fixed space to regulate our spacing in other ways within the home. We sit at the same place at the table, and many sit in the same chair to relax. This type of behaviour organises much of our lives. We use the same principles in other groups when we have separate offices or workrooms, using walls to maintain separation between people.

Most families tend to become crowded as the children reach sexual maturity. Most parents still extrude the young at this stage, generally after a period of varying degrees of disruption from agonistic behaviour between generations. The problem is accentuated by emphasis on higher education. The basis of parental behaviour is the use of resources by adults to ensure successful growth and reproduction of their young. Education is one key to success of youngsters, and parents use their resources to support their young through a long education process. Yet the retention of dependent offspring well into adult life does generate considerable stress in the majority of families. One of the most common forms of social adaptation here is to send the youngsters away to finish their higher education.

The provision of privacy within the home or at work is a recent development. Walls and rooms are expensive, but are the most satisfactory way of controlling personal space that we have yet discovered. Animals also use this system when they use natural barriers as territorial boundaries. We have the ability to structure the environment at will. Amongst other things, we use the privacy gained to recover from periods of stress. Privacy is probably only necessary in large anonymous societies.

The control of space in buildings and rooms is always expensive. All can benefit from the provision of adequate private space, but all cannot afford the cost. The distribution of resources in our societies remains as unequal as it is in most other species. High social status brings rewards which are reinforcing to the competitiveness of those who are ambitious. It seems important that ambition and service should be rewarded since they provide much of the drive which develops our societies. Yet we have also made many social changes to ensure that the distribution of wealth is not solely dependent on competition. We have pensions of many types, insurance, with opportunities to own houses supported by governments, while trade unions bargain for an adequate share of the wealth produced. Yet it does seem that we must go much further, for the crowding of families can only be eliminated if the vast wealth now produced is distributed to those now unable to secure adequate living space.

In conclusion, we do have the resources to control most forms of crowding. It may be possible to adapt by changing our nature or the nature of our societies, but it seems simpler to adapt our environment. This can occur by organisation of our societies to direct resources to the problem. Some forms of crowding appear difficult, particularly the transport problem. Yet this must also be solved by provision of money, for we are not likely to give up the satisfying life we have developed by our complex group structure.

Summary

Some of the natural subdivisions of animal societies have been described in terms of the spacing relationships between individuals. The most significant are the aggregations, the caste structure of the animals, and the functional organisation of societies in time. These factors contribute to the density of animal populations and thus to a definition of crowding. Animals occupy two types of space: that actually displaced by the animal and a personal area around it. The personal areas vary in several ways, but crowding develops when animals are forced to enter the personal areas of neighbours.

An attempt has been made to present a comparative approach to the study of human and animal societies, with a discussion of the evolution of human societies. The comparative approach was then extended to questions of crowding in modern human societies.

References

1 G. McBride, 'Society Evolution', *Proc.ecol.Soc.Aust.*, **1**, 1966, 1-13.

2 G. H. Orians and M. F. Willson, 'Interspecific Territories in Birds', *Ecology*, **45**, 1964, 736-45.

3 H. P. Hediger, 'The Evolution of Territorial Behaviour' in *The Social Life of Early Man*, S. L. Washburn (ed.), London, 1962, Methuen, 17-33.

4 H. Hediger, *Wild Animals in Captivity*, London, 1950, Butterworth.

5 G. McBride, 'A General Theory of Social Organization and Behaviour', *Univ. of Queensland Vet.Sci.Pap.*, **1**, 1964, 75-110.

6 G. McBride and F. Foenander, 'Territorial Behaviour in a Flock of Domestic Hens', *Nature* (London), **194**, 1962, 102.

7 G. McBride, 'Social Organization and Stress in Animal Management', *Proc.ecol. Soc.Aust.*, **3**, 1968, 133-8.

8 G. McBride, J. W. James, and R. N. Shoffner, 'Social Forces Determining Spacing and Head Orientation in Domestic Hens', *Nature* (London), **197**, 1963, 1272-3.

9 G. McBride, 'Social Organization and Stress in Animal Management', *Proc.ecol. Soc.Aust.*, **3**, 1968, 133-8.

10 H. H. Shoemaker, 'Social Hierarchy in Flocks of Canary', *Auk*, **56**, 1939, 381-405.

11 R. S. Palmer, 'A Behaviour Study of the Common Tern', *Proc.Boston Soc.Nat. Hist.*, **42**, 1941, 1-119.

12 C. H. Southwick, 'Social Behaviour of Rhesus Monkeys in a Temple Habitat in Northern India', *Amer.Zool.*, **1**, 1961, 262.

13 G. McBride and F. Foenander, 'Territorial Behaviour in a Flock of Domestic Hens', *Nature* (London), **194**, 1962, 102.

14 J. L. Cloudsley Thompson, *Animal Behaviour*, Edinburgh, 1960, Oliver and Boyd.

15 C. H. Southwick, 'Social Behaviour of Rhesus Monkeys in a Temple Habitat in Northern India', *Amer.Zool.*, **1**, 1961, 262.

16 G. McBride, 'Social Organization and Stress in Animal Management', *Proc. ecol.Soc.Aust.*, **3**, 1968, 133-8.

17 G. McBride, 'Society Evolution', *Proc. ecol.Soc.Aust.*, **1**, 1966, 1-13.

18 K. Lorenz, *King Solomon's Ring*, London, 1952, Methuen.

19 G. McBride, 'On the Evolution of Human Language', *Social Science Information*, **7**, 1969, 81-5.

20 R. S. Palmer, 'A Behaviour Study of the Common Tern', *Proc.Boston Soc.Nat. Hist.*, **42**, 1941, 1-119.

Comments

J. D. FREEMAN First, let me express my appreciation of Dr McBride's paper with its wealth of facts and theory. As an anthropologist with an interest in ethology I am in sympathy with his comparative and evolutionary approach to the study of behaviour, but I must disclaim any kind of expert knowledge of the phenomenon of crowding. All I can do is to indicate a few points that would seem to warrant further investigation in this little-understood field. Given the theme of this symposium I shall confine my attention to behavioural aspects of crowding in the human species; and further, because of the limitations of time, I shall not comment at any length on Dr McBride's thought-provoking hypothetical reconstruction of hominid societal evolution —which is material enough for another entire symposium.

Here, a principal issue is the phylogenetic antiquity of the hominid pair-bond. In Dr McBride's scheme it is proposed that large-scale hominid groups evolved by way of the 'federation of families'. Another very different hypothesis, advanced by Reynolds and others,[1] accepts the existence (in the Pliocene and earlier) of open groups (comparable to those to be found among chimpanzees) and envisages the emergence of the pair-bond and the home-base as evolutionary adaptations to savannah-based predation. These and other reconstructions of the possible course of hominid evolution are indications of the presently prevalent reawakened interest in human origins which springs from the realisation that man's nature cannot be adequately understood other than in evolutionary terms. The full course of hominid evolutionary history will never be recovered, but this does not negate the value of making hypothetical interpretations of the evidence now available and subjecting them to inferential testing, and it will be of interest to examine Dr McBride's set of interpretations when they are published in more detail in book form. At this rudimentary stage in the study of hominid evolution any hypothesis that stimulates thinking about fundamental issues is to be welcomed.

Crowding occurs, Dr McBride has said, when animals are forced to enter the personal areas of conspecifics, and he has stressed the importance of spacing in the social behaviour of humans, as in other species. Ironically, our knowledge of spacing in birds and a variety of other animals is more advanced than with our own species, but there is evidence enough to indicate that comparable mechanisms are integral to human social behaviour. That notable *Homo sapiens*, W. H. Auden,[2] for example, has recently expressed his feeling about personal spacing, in these splendidly mettlesome lines:

> Some thirty inches from my nose,
> The frontier of my Person goes,
> And all the untilled air between
> Is private *pagus* or demesne.
> Stranger, unless with bedroom eyes
> I beckon you to fraternise,
> Beware of rudely crossing it:
> I have no gun, but I can spit.

In the Samoan Islands (of Western Polynesia) where I recently devoted two years to the observational study of behaviour under natural conditions, spacing is a prime indicator of social rank. For example, during the ritualised serving of *Kava*, the distance which the cup-bearer withdraws before facing and then approaching a particular chief is a conspicuous indication of his rank relative to the other chiefs who happen to be present. Further, in informal situations, as, for example, when a chief of high rank happens to be standing and talking to a number of titled men of lower rank, the greater extent of his personal space is clearly discernible.

In this case those of lower rank, while engaged in friendly discourse with him, are displaying some degree of deference to an individual whom they recognise as having higher rank than themselves; and, as I would interpret such behaviour, it is a human refinement of the spatial equilibration characteristic of many infra-human primates, and which has been well described for *Macaca mulatta* by Chance.[3]

The prime adaptive function of equilibration is the avoidance of contentious situations and the maintenance of social homeostasis; and it would seem that, with the phylogenetically-given behaviours which produce equilibration, mesolithic man was, in important ways, pre-adapted to the urban aggregations which have since emerged in the course of human history. Nonetheless, there must be limits to the degree of crowding which humans can sustain without pathological consequences. As the nightmarish experiments of Calhoun[4] have indicated, extreme crowding in animals can culminate in a drastic disintegration of behaviour and in gross physiological disturbances. Here we are confronted with the severe terminal phase of a pathological process; but there is no reason to suppose that the onset of these disturbances is not gradual and that incipient and not easily detectable effects are present at lower densities of crowding. In the human situation it is these possible incipient effects that deserve to be studied in detail so that we may stop well short of 'the behavioural sink'.

Dr McBride has referred to the houses and apartments in which we live as nests. This they may sometimes be, but with their walled enclosures and lockable doors they may also become cages. In Samoa, young children are not infrequently subjected to severe and hurting parental punishment which often reduces them to a distraught emotional state. So severe is the régime that one would expect serious psychological consequences. Some consequences there certainly are—but not of the expected order. When I came to investigate this situation more closely, my observations showed that after a child had been punished he was left alone, and, as Samoan houses have no walls whatsoever he was soon able to *withdraw* and seek the company of other children somewhere nearby. With this behavioural pathway open to him a child was able to move from one mood to another in a matter of ten or so minutes; that is, from his distraught and resentful mood after punishment to the quite different mood of play behaviour. When I had plotted this behavioural sequence it occurred to me how different the outcome would have been if the child in question had been *confined* by parental decree to his room in a European-style house, for then there would have been no rapid mood shift, and the child's resentment would have persisted with possibly paranoid consequences. I mention this because it seems to me that one of the important tasks of urban biology is the detailed investigation of the behavioural consequences of existence in our cities—and particularly in their amorphous suburbs with their serried rows of ticky-tacky boxes.

Yesterday, Professor Hetzel drew attention to what he called 'the impersonal urban situation'. When one returns, as I have recently done, from a prolonged period of living in a natural kin-based grouping of some hundreds of people with whom one has diverse and genuinely emotional relationships hour by hour throughout the day, one senses the peculiar paucity of the suburb as a social environment, and one suffers, indeed, from a kind of cultural shock which is essentially a realisation of the degree of behavioural and emotional deprivation involved. Throughout almost the whole of their evolutionary history men have flourished in natural groupings comparable to the kind I have just described, for even ancient cities had their long-established neighbourhoods. The suburbs are something different—the often unsightly spawn of our technological centres—and a markedly novel social environment which—as Professor Hetzel has suggested—may well have consequences of a subtle yet ultimately devastating kind for some of the people who live in it, and fail to adapt to its unusual demands. I have touched briefly on this topic because it seems to me that in some circumstances dispersal is perhaps more unhuman than crowding.

Next, may I direct attention to an aspect of crowding which would seem to deserve mention on such an occasion as this. I refer to the formation in large cities of crowds that subsequently become violent and destructive mobs. Mob behaviour is, unquestionably, one of the most conspicuous phenomena of our age, and a major police problem, especially in countries with urban populations—but the aetiology of mob behaviour remains very imperfectly understood. Colonel Rex Applegate,[5] a leading American authority on crowd and riot control, has recently given this account of the way in which a spontaneous crowd becomes a violent mob. 'Its members', he writes, 'lose their identity as individuals and merge into a cruel and primitive body which has lost civilised restraints and suddenly has no respect for law and order'. This language may seem somewhat melodramatic—but these are the terms in which straight-faced police officers write of mob behaviour in their technical manuals. Ethological research has given us some understanding of the way in which, in a crowd, mood transmission is effected by way of behavioural mechanisms, but we still have to account for the behavioural subphase, to use Dr McBride's terminology, into which a violent mob does shift.

We are dealing here with a recurring aspect of human social behaviour, and I would like to ask Dr McBride whether he considers violent mob behaviour to be the result of crowding, or whether he would propose some other kind of aetiology? My own view is that there are good grounds for the investigation of the phylogenetic basis of such behaviour.

Finally, may I make two brief comments on the general theme of this symposium. From the research on which I am currently engaged on the history of ideas about human behaviour, I have become persuaded that one of the notable effects of civilisation on the biology of man has been the proliferation of erroneous views as to his nature. This was conspicuous in the doctrines of Rousseau and the ideologues, persisted—despite Darwin—through the nineteenth century, and, in the first quarter of this century, led to a disastrous splitting-off of the social from the biological sciences. It is heartening that in recent years a new intellectual mood has developed, and that in symposia such as this present one, an integrated scientific approach to *bio-cultural man* is beginning to emerge.

Last of all, I feel that at one point in this symposium, however briefly, homage should be made to the *truly* remarkable achievements of human civilisation—ranging, as they do, from such things as the Bodhisattvas of the East to 'The Magic Flute' in the West. The more these bits of civilisation make an impact on my 'biology', the better I am pleased; and whatever the pollution of our cities, I would hope with W. H. Auden[6] (and I am sure, also, Professor Dubos)

> That crowds two centuries from now will press
> (Absurd their hair, ridiculous their dress)
> And pay in currencies, however weird,
> To hear *Sarastro* booming through his beard.

[1] V. Reynolds, 'Open Groups in Hominid Evolution', *Man*, 1, 1966, 441-52.
[2] W. H. Auden, *About the House*, London, 1966, Faber and Faber. Reproduced with the permission of the publishers.
[3] M. R. A. Chance, 'Social Structure of a Colony of Macaca mulatta', *Brit.J. of Animal Behaviour*, 4, 1956, 1-13.
[4] J. B. Calhoun, 'A "Behavioural Sink"', in *Roots of Behaviour*, E. L. Bliss (ed.), New York, 1962, Harper and Bros., 295-315.
[5] R. Applegate, *Crown and Riot Control*, Harrisburg, Pa., 1964, Stackpole.
[6] W. H. Auden, 'Metalogue to The Magic Flute', in *Collected Shorter Poems*, 1927-1957, London, 1966, Faber and Faber. Reproduced with the permission of publishers.

Comments

K. MYERS Crowding in laboratory and natural populations of many mammals causes significant changes in behaviour and physiology. The kinds of changes which occur are indicated by data from adult male rabbits in experimental populations.[1] (See Table 6:1. All the data in this and the following tables are corrected for regression on age.)

There is a large loss in body weight, and changes in the sizes and condition of index organs vitally concerned with metabolic function (spleen, liver, kidneys). The kidneys become inflamed and scarred with lesions of a systemic disease similar to that measured in mice and other mammals, where it has been shown to be ACTH-induced.[2] Reproduction is impaired, and the adrenal glands hypertrophy and lose their lipid steroid precursors. The zonation of the adrenal alters to favour increased secretion of glucocorticoids and suppression of mineralocorticoids, possibly increasing natriuresis and upsetting Na : K ratios in the body. Adrenal nodulation (the formation of small adenomas budded off by the adrenal cortex and consisting largely of

Table 6:1

Relationships between indices of condition, behaviour, and mortality in male rabbits

Variable	Type of death				Significance of Differences between means
	Sacrificed at end of experiment	Social persecution	Pathological symptoms	Unknown	Deg. fr. 3/127
n	87	17	23	3	
Survival (per cent)	100	50·5	63·4	64·2	$p < \cdot 001$
Weight at death (g)	1469	1156	1150	1066	$p < \cdot 001$
Change of weight (per cent)	—6·1	—24·9	—26·1	—29·8	$p < \cdot 001$
Relative adrenal weight (g/kg)	0·305	0·527	0·502	0·612	$p < \cdot 001$
Lipid in adrenal *Z. fasciculata-reticularis* (Index 1-10)	9·07	5·18	5·44	4·25	$p < \cdot 001$
Total area of median cross-section of adrenal (mm^2)	31·56	40·48	36·62	40·63	$p < \cdot 001$
Area of adrenal *Z. glomerulosa* (per cent total area)	10·2	7·0	7·0	7·5	$p < \cdot 001$
Area of adrenal *Z. fasciculata-reticularis* (per cent total area)	83·9	87·5	87·7	89·1	$p < \cdot 001$
Numbers of nodules on adrenals (per rabbit)	1·25	0·41	0·44	0·00	$p < \cdot 01$
Alveolation of adrenal *Z. fasciculata-reticularis* (Index 1-12)	7·44	10·12	10·35	8·50	$p < \cdot 01$
White blood cells in adrenal cortex and medulla (Index 1-9)	1·03	4·65	5·13	4·25	$p < \cdot 001$
Pituitary weight (mg)	4·42	—	2·25	—	$p < \cdot 001$
Social status (1 = dominant)	2·05	2·74	2·04	2·35	$p < \cdot 05$
Total aggressive behaviour (acts/minute)	0·055	0·035	0·070	·017	$p < \cdot 01$
Total sex behaviour (acts/minute)	0·079	0·020	0·073	0·016	$p < \cdot 05$

Myers and Hale (unpublished).

glomerulosal tissue)[3] is suppressed, thus exacerbating problems relating to Na and K metabolism already posed by pituitary secretion of ACTH. The adrenal cortex is invaded by small, undifferentiated white blood cells, which appear first in the medulla and then throughout the cortex against the direction of blood flow, and probably represent hematopoietic activity by the adrenals themselves.[4] The adrenal cortical tissue is reorganised into alveoli separated by thin-walled sinuses. Parallel changes occur in the human adrenal[5] when abundant steroidogenesis occurs. Similar structural change is also described in adrenals of other stressed mammals–*Microtus, Cervus*.[6]

Recent experiments (Myers *et al.*, unpublished) show that the key to reaction to density lies not in increase in numbers *per se*, but to some quality of decrease in living space, which may well be intrusion by other members of the group into personal space, discussion of which forms a large and important part of Dr McBride's paper.

Heightened activity of the pituitary-adrenocortical system during emotional stress has also been demonstrated in several studies of man. During competitive sports, anticipation of surgery, exposure to shame, and emotional disturbances of various kinds, especially where fear of physical injury is involved, the adrenal cortical hormone level in the plasma rises and increased amounts of 17-hydroxycorticosteroids are excreted in the urine. When distress passes the circulating corticoids decrease substantially.[7]

The effects of stressful factors in rabbit populations are also evident at embryonic, nestling, and young animal level. Nestlings born under conditions of high density exhibit a marked retardation in development of all body proportions.[8]

There are also significant differences in zonation and morphology in the young adrenals, which point to large differences in types and rates of secretion of cortical hormones.

Table 6:2

Effects of population density on nestling rabbits

Variable	High density	Medium density	Low density	Significance of differences between means Deg. fr. 2/143
Body wt (g)	113·3	136·2	147·8	p<·001
Tarsus (mm)	37·9	38·1	44·9	p<·001
Shoulder fat (g)	0·52	0·72	0·76	p<·001
Thymus (g)	0·25	0·32	0·32	p<·001
Adrenals (g)	0·012	0·012	0·016	p<·001
Liver (g)	4·56	5·39	7·35	p<·001
Kidney (g)	1·43	—	2·19	p<·001
Spleen (g)	0·093	0·087	0·128	p<·001
n	97	42	7	

Myers and Mykytowycz (unpublished).

Table 6:3

Effects of population density on weaned young

Variable	Males			Females			Analysis of variance.	
	High density	Low density	Significance of differences between means Deg. fr. 1/18	High density	Low density	Significance of differences between means Deg. fr. 1/20	Sex	Treatment
Body weight (g)	344	393	p< ·05	398	454	p< ·1	N.S.	p< ·05
Adrenals (g)	0·032	0·044	p< ·05	0·037	0·061	p< ·001	p< ·05	p< ·001
Total area section (mm^2)	5·80	6·56	N.S.	4·93	6·56	p< ·001	N.S.	p< ·05
Area *Z. fasciculata-reticularis* (per cent total area)	70·9	64·7	p< ·001	69·6	65·2	p< ·01	N.S.	p< ·05
Area medulla (per cent total)	13·4	18·6	p< ·001	16·6	19·1	p< ·001	N.S.	p< ·05
Alveolation *Z. fasciculata-reticularis* (Index 1-12)	5·88	3·63	p< ·001	5·86	3·85	p< ·001	N.S.	p< ·05
n	9	10		11	10			

Myers and Hale (unpublished).

Table 6:4

Relationships between origin of male rabbits and their biology as adults

Variable	Origin of rabbits				Significance of differences between means Deg. fr. 3/127
	Born in high density confined population	Born in medium density confined population	Born in low density confined population	Born in natural population	
Survival (per cent)	75·8	68·3	84·4	86·6	p< ·1
Body weight (g)	1200	1225	1389	1375	p< ·1
Change of weight (per cent)	—21·0	—22·0	—9·1	—12·0	p< ·05
Relative adrenal weight (g/kg)	0·421	0·483	0·315	0·363	p< ·1
Lipid in adrenal *Z. fasciculata-reticularis* (Index 1-12)	5·67	5·80	8·33	8·07	p< ·001
Area *Z. glomerulosa* (per cent total area section)	8·7	8·2	9·2	10·3	N.S. but climbing
White blood cells in medulla and cortex (Index 1-9)	3·67	4·40	2·42	2·04	p< ·01
Numbers of nodules on adrenals (per rabbit)	0·17	0·40	0·83	1·08	N.S.
Alveolation of *Z. fasciculata-reticularis* (Index 1-12)	10·17	9·20	9·42	8·01	N.S. but climbing
Status (1 = dominant)	1·83	2·10	1·67	2·25	N.S.
Total aggressive behaviour (acts/minute)	0·122	0·082	0·072	0·039	p< ·1
Total sex behaviour (acts/minute)	0·136	0·069	0·136	0·083	N.S.
Total activities (per minute)	0·311	0·210	0·227	0·173	p< ·01
Testis weight (mg)	2045	1846	1944	1615	N.S.
Stage of spermatogenesis	8·0	7·63	8·0	7·51	N.S.
n	6	10	12	102	

Myers and Hale (unpublished).

Male rabbits born under such crowded conditions are very active as adults, showing high rates of aggressive and sexual activities. They have large adrenal glands, which give evidence of heavy rates of secretion, and show large losses in body weight and a decided tendency towards lower rates of survival. Female rabbits born under crowded conditions show less significant differences, but still possess heavy adrenals with apparently heightened secretory activity, and exhibit lower survival rates. Phenomena present in crowded populations thus appear to be capable of predetermining patterns of behaviour, health, and survival in adults by effects on them when young. It is critically important to know whether the causal mechanisms are predominantly genetic, sociological, or physiological in nature.

A genetic mechanism is not implausible. In a recent, unpublished study, Dr Charles Krebs of Indiana University, Blomington has shown that in voles, *Microtus spp.*, there are highly significant changes in aggressive behaviour during population cycles, with the most aggressive animals occurring in peak populations. Krebs studied transferrin polymorphism in the blood over a cycle, and measured changes in gene frequency of nearly 50 per cent over six weeks—a rate of change hitherto unsuspected in mammalian populations. On the other hand the psychological and physiological literature is rich in data showing that endocrine imbalance in pregnant female mammal and embryo, lactating female, and neonatal young, causes significant changes in adult behaviour and physiology.[9] It seems likely that hormones directly affect the central nervous system during development to produce permanent changes in physiology and behaviour in later life.[10] A voluminous literature testifies to the fact that environmental and social factors during infancy are more profound in their effects than during any other period of mammalian life.

Man is not divorced from these classes of events. Genetically controlled abnormalities of adrenocortical hormone secretion and metabolism are relatively common in the human population. The most common, congenital adrenal hyperplasia, causes masculinisation in girls and precocious puberty in boys, and is due to an autosomal recessive gene with an estimated frequency in heterozygotes from 1 in 35 in Switzerland (1958) to 1 in 125 in Maryland (1956).[11] Developmental abnormalities of important kinds caused by physical and social environmental stimuli also occur. Thus emotional state in the human female, physical agents, nutrition, drugs, maternal diseases, and age have all been shown to affect the developing human foetus[12] leading to variations in congenital malformations,[13] neo-natal death rates and birth weights[14] and seasonality in the births of the mentally deficient.[15]

My aim in introducing discussion on Dr McBride's paper in this fashion is twofold. Firstly, to intimate that modern work in mammal ecology supports the kind of thing that he is talking about, and secondly to emphasise that behaviour is not something to be studied on its own. There is an important organic basis to behaviour and physiology, which unfortunately tends to be forgotten in many modern sociological, psychological, and anthropological areas of thinking. This organic component has its roots in developmental homeostasis, and crowding is one of the most potent of the complex stimuli affecting it. There is no doubt in my mind that these phenomena rank importantly in the life of man.

No manner of specious thinking can separate the human mammal from the biological basis of his existence.

[1] J. J. Christian, J. A. Lloyd, and D. E. Davis, 'The Role of Endocrines in the Self-Regulation of Mammalian Populations', in *Recent Progr. in Hormone Research*, New York, 1965, Academic Press, 501-78; K. Myers, 'The Effects of Density on Sociality and Health in Mammals', *Proc.Ecol.Soc.Aust.*, **1**, 1966, 40-64; D. Thiessen, 'Population Density and Behaviour: a review of Theoretical and Physiological Contributions', *Texas Rpts on Biol. and Med.*, **22**, 1964, 266-314.

[2] J. J. Christian, 'ACTH—Induced Renal Glomerular Disease in Intact, Adrenalectomized and Castrate Male Mice', *Proc.Soc.Exper.Biol.Med.*, **126**, 1967, 152-7.

[3] K. Myers, 'Morphological Changes in the Adrenal Glands of Wild Rabbits', *Nature* (London), **213**, 1967, 147-50.

[4] H. W. Deane, 'The Anatomy, Chemistry and Physiology of Adrenocortical Tissue' in *Handbuch Der Experimentellen Pharmakologie Teil 1—The Adrenocortical Hormones*, H. W. Deane (ed.), Berlin, 1962, Springer Verlag, 23.

[5] T. Symington, 'Morphology and Secretory Cytology of the Human Adrenal Cortex', *Brit.Med.Bull.*, **18**, 1962, 117-21.

[6] J. J. Christian and D. E. Davis, 'Endocrines, Behaviour and Population', *Science*, **146**, 1964, 1550-60; J. J. Christian, 'The Pathology of Overpopulation', *Military Medicine*, **128**, 1963, 571-603.

[7] L. Levi, 'Endocrine Reactions during Emotional Stress', in *Emotional Stress*, L. Levi (ed.), New York, 1968, Elsevier Pub. Co., pp. 109-13.

[8] J. J. Christian, J. A. Lloyd, and D. E. Davis, 'The Role of Endocrines in the Self-Regulation of Mammalian Populations', in *Recent Progress in Hormone Research*, New York, 1965, Academic Press, 501-78.

[9] V. Denenberg (in press), 'The Effects of Early Experience', in *The Behaviour of Domestic Animals*, 2nd ed., E. S. E. Hafez (ed.), London, 1969, Ballière, Tindall and Cox.

[10] W. C. Young, 'The Hormones and Mating Behaviour', in *Sex and Internal Secretions*, W. C. Young (ed.), Baltimore, 1961, Williams and Wilkins, 1173-239.

[11] D. A. Hamburg, 'Genetics of Adrenocortical Hormone Metabolism in Relation to Psychological Stress', in *Behaviour—Genetic Analysis*, J. Hirsch (ed.), New York, 1967, McGraw-Hill, 154-75.

[12] M. F. A. Montagu, 'Constitutional and Prenatal Factors in Infant and Child Health', in *Human Development*, M. L. Haimowitz and N. R. Haimowitz (eds.), New York, 1960, Crowell, 126-42; I. W. Monie, 'Influence of the Environment on the Unborn', in *Human Ecology*, J. B. Bresler (ed.), Reading, Mass., 1966, Addison-Wesley, 303-11.

[13] J. R. Gentry, E. Parkhurst, and G. V. Bulin, 'An Epidemiological Study of Congenital Malformations in New York State', in *Human Ecology*, J. B. Bresler (ed.), Reading, Mass., 1966, Addison-Wesley, 223-47.

[14] D. Grahn and K. Kratchman, 'Variation in Neonatal Death Rate and Birth Weight in the United States and Possible Relations to Environmental Radiation, Geology and Altitude', in *Human Ecology*, J. B. Bresler (ed.), Reading, Mass., 1966, Addison-Wesley, 251-76.

[15] Hilda Knobloch and Benjamin Pasmanick, 'Seasonal Variation in the Births of the Mentally Deficient,' *Environments of Man*, J. B. Bresler (ed.), Reading, Mass., 1968, Addison-Wesley, 157.

Discussion

SIMONS The discussion of the effect of over-crowding as we have seen is usually concerned with occupancy of physical space and physical proximity; that is, of physical territory, and of social territory. I am almost completely unfamiliar with this field, but I wonder whether at least in the case of man there might be a third category. This category, it seems to me, has elements which are distinct from social territory and which include important determining factors in the response to physical space. For example, it is conceivable that some degree of physical confinement may not engender mental stress, and conversely, that while in virtual physical isolation a person may feel that his mental territory is being encroached upon and his stake to an area of mental territory being challenged. Again, the claim that the basis for a person's views are non-rational may be seen as a challenge to that person's thinking patterns within his particular mental territory. My question is: Is it possible that these sorts of situations have aspects that are different from the concept of social territory or is it usual to incorporate them within that schema?

LEMBERG One difficulty I see is that in humans the requirement for the individual space of an individual may be entirely different from one to another individual, much more so than in other animals.

MCBRIDE Of course there is a large amount of variability in the amount of space controlled by animals just as there is in man. The important thing is, of course, that this is an almost universal feature of behaviour in animals and man. It is learned, perhaps without using the typical human verbalising approach. We acquire this in our growth without somebody explaining that it is some thirty inches in front, or it is seventeen inches at the side, and you can actually touch shoulders.

Regarding the fact that we often seek crowding and high levels of interaction, I have always argued that crowding involves this intrusion into personal space and this is always alerting. Mr Myers has picked up the next effect at the physiological level; I have tried to keep only to the social level. There appears to be an optimum amount of social contact required and this is probably true of every species. You will see an animal alert to the lack of social contact just as much as it will alert to too much. And, of course, in parties and such situations we go out specifically to create a subphase with certain properties to engage in a large number of interactions. But we cannot keep this up indefinitely nor do we ever try. So this idea of an optimum leads to the seeking out of special situations of high and low interaction frequency. Of course, we not only have an optimum number of interactions, but we also have an optimum number of each type of interaction. For the affiliative type of interactions, we almost certainly have an optimum here. It is true that in most animal groups, and in primitive societies, individuals are easily able to habituate or require a very high frequency of affiliative interactions with individuals with which they are permanently in association and have very strong affiliative bonds. Of course one of the problems of our modern type of society is that we form affiliations specified in work groups, in play groups, in family groups. In all these different types of groups that we jump between and clutter our roads to do so, we have affiliations at various levels, and most of us are able to get whatever stimulation we need from affiliations in these. Some individuals are unable to acquire enough of these affiliative relationships and suffer from deficiencies here, and this is certainly a field that we need to know a lot more about. The phenomenon of crowding in large anonymous groups—this of course is a phenomenon absent in man's evolutionary history. In many of our groups, entertainment groups, cinemas, dances and so forth, we are always coming into reasonably anonymous groups, so that we have a large repertoire of appropriate behaviour. If we intrude upon strangers, we have standard forms of interactions, such as apologies, which are adaptations to this sort of crowding, and we know how to deal with them. But mob behaviour is quite different. We do have something analagous to this in animals, though not the aggressive type. The hen rears her broods of chickens and keeps them totally isolated for the whole of the time she protects them. For about twelve weeks they never come in contact with any other individual but the members of their brood and their dam. She finally drives them off and returns to the male. They then live a life in this tiny little group moving over the same area, doing the same things at the same times each day that they did with her. As they start to approach sexual maturity a new change occurs. These groups have to be federated into one group before they move into the adult flock as sexual maturity is completed. And they do this by streaming; one individual runs and all these birds run and join in and a great stream of birds appears running across an open field, running somewhere, quite specifically to a tree that they know. Now, this is happening at about eighteen weeks of age. In large fowl-houses where many thousands of birds are aggregated together, at this age a phenomenon called 'fowl hysteria' sometimes develops. A bird runs and other birds join. They start running but have nowhere to run for it is an unstructured environment. They run round and round, they build up the sawdust into an arena as they do so and then they finally stop from exhaustion, and twenty minutes later they start it up again in exactly the same way as the fowls do in the wild. Thus, when you take behaviour that is perfectly appropriate in one set of circumstances and put it into another, it can become quite abnormal and can be quite destructive to the individual.

7 Cedric Mims

Stress in Relation to the Processes of Civilisation

Even Hans Selye, the pioneer of the stress concept, had difficulty in defining stress.[1] But words can be worth using, even when they are hard to define. In the popular sense, the word stress often refers to *any* noxious stimulus or situation, and sometimes indeed to any stimulus to which the organism responds. I shall use stress in a more restricted sense. Stress in the first place is any disturbance in body homeostasis general enough or severe enough to call into action a co-ordinated bodily response—a response which has been designed in evolution to cope with such disturbances. These disturbances include cold, hunger, physical exertion, infection, and injury, and in each case the response tends to involve the same set of hormonal and nervous mechanisms.

One should distinguish the stress stimulus (stressor) which induces the state of stress, from the response of the organism to this stress. Also, it is convenient to distinguish physical stress stimuli from mental stress stimuli. It is in the central nervous system that the response to a physical stress originates. The central nervous system can also initiate these changes without the actual stimulus of physical stress, and this may be necessary, for instance, in the immediate preparation for running or fighting. But the changes also occur in response to or in preparation for a stress that is social or psychological rather than physical. In other words, the stress is one that disturbs social or psychological homeostasis. The changes nevertheless tend to be the ones appropriate for dealing with a physical stress. In human beings, indeed, they occur in response to a stress which is only threatened, imagined, or symbolic. Such mental stresses have come to be more and more important in human evolution and it is with them that I shall be dealing most of the time.

MECHANISMS IN THE RESPONSE TO STRESS

It is worth briefly describing the mechanisms by which the stress responses operate, to understand how generalised they are, and, at the cost of some simplification, to see their adaptive significance.

The bodily changes of stress originate from the central nervous system, more specifically from the ancient limbic system, acting via the hypothalamus. The hypothalamus in turn acts on the autonomic nervous system and on certain endocrine glands (see Fig. 7:1).

Autonomic nervous system

Stress changes mediated by the autonomic nervous system have been thoroughly documented, and Cannon[2] showed how they can be interpreted as preparation for appropriate bodily action, such as fight or flight. From the hypothalamus messages travel down autonomic nerves to muscles and blood vessels, and also to the adrenal medulla where catecholamines (epinephrine and norepinephrine) are released into the blood stream. As a result, the heart beats faster and pumps out more blood, the blood vessels in muscle dilate, digestive functions are depressed, the blood clots more readily, glucose is released into the circulation and one feels alert, prepared. This is what happens for instance in competitive sports, or in parachute jumping.[3] Since mental

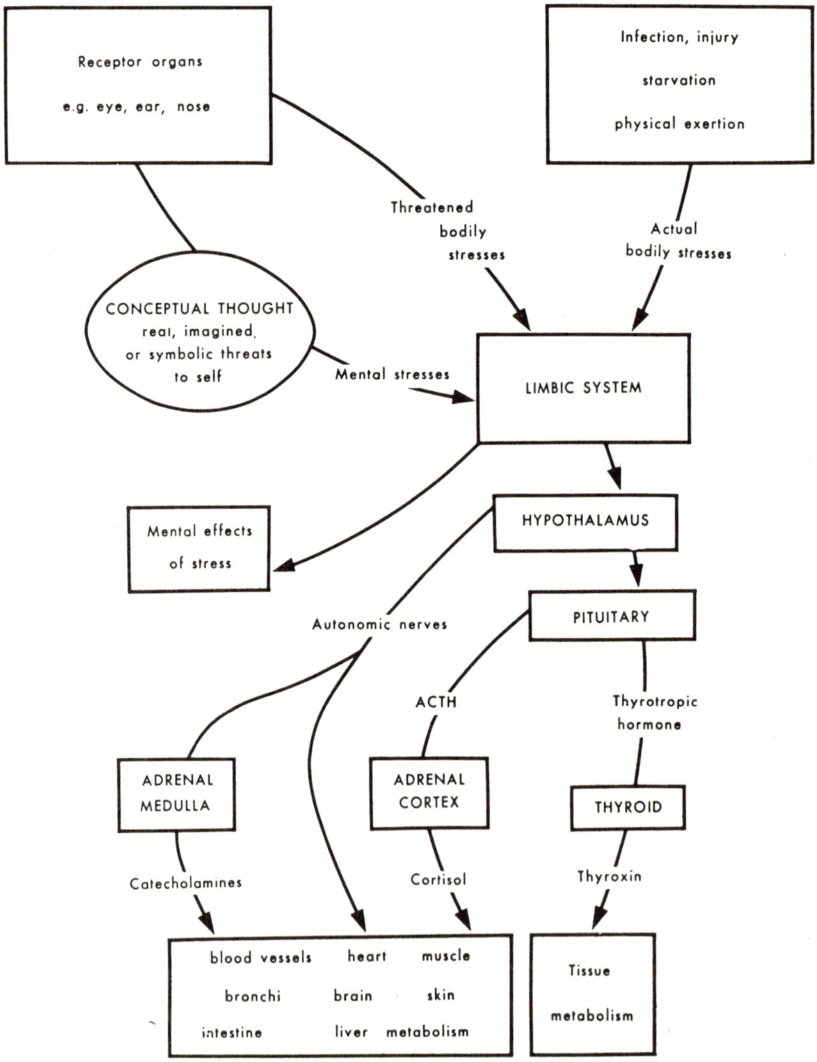

Fig. 7:1 Diagrammatic representation of stress mechanism in man

stress or excitement by itself can produce the changes that prepare the body for action, changes of this type also occur at stressful interviews or in viewing certain emotion-laden movies.[3] The thought or the threat of a metaphorical slap on the face may be as effective as a real slap on the face.

The classical work on stress changes mediated by the autonomic nervous system was done with Tom, the man with a gastric fistula.[4] The physicians who employed Tom as a laboratory assistant were able to study directly the changes in vascularity, secretion, and motility of his gastric mucosa. When he was frightened or depressed his gastric mucosa blanched, like his face, and secreted less acid; when he was angry his gastric mucosa became red, as did his face, and secreted more acid. The state of Tom's mind profoundly influenced the state of his gastric mucosa, and much was thus learnt about the mechanism of production of peptic ulcers. Since then, similar studies have been made of human rectal mucosa and even small intestinal mucosa, with comparable findings.

There are many other changes mediated by the autonomic nervous system during mental stress. For instance, one useful physical indication of anxiety is an increase in forearm blood flow.[5] There are also the stresses of misery or hopelessness as well as those of fight, flight, and fright. The precise pattern of change depends both on the individual and on the stress stimulus.

Adrenal cortex

In the stress response, the message from the hypothalamus also goes to the anterior pituitary gland, which then produces more ACTH (adrenocorticotrophic hormone). The ACTH acts on the adrenal cortex, increasing its output of hormones. The glucocorticoid hormones are of great importance in the response to stress. Perhaps the best way to understand their action is to think of local bodily reactions as being predominantly vascular, with dilation of small blood vessels or inflammation. When local, these responses are excellent and appropriate, but if they were to take place on too large a scale, all over the body, as in general muscular activity, infection, or serious injury, it would be different. If each part now responded as vigorously as to a single local stimulus, the circulation would collapse, with disastrous results.

The glucocorticoid hormones, however, inhibit these vascular responses, mobilise glucose and free fatty acids from stored protein and fat, and allow catecholamines to have certain important actions. As a result, general bodily functions are kept intact so that there can be a generalised and appropriate response. For the moment, in fact, the parts are sacrificed for the good of the whole. This simplified concept of the action of adrenal cortical hormones dampening down local responsiveness for the general bodily good was foreshadowed by Selye[6] who referred to 'the fire alarm system which puts out the flame of inflammation'. More recently it has been extended and developed in some detail by Schayer.[7]

The adrenal cortical response takes place particularly during the physical stress of vigorous muscular activity, starvation, injury, surgical operation, or infection. Like the autonomic nervous system response to physical stress, it has been of immense survival value during evolution, and its importance is illustrated in the clinical picture seen in patients with adrenal cortical insufficiency. In these patients the commonest symptom of all is general asthenia (weakness); this is first experienced during times of stress, and it later becomes continuous, requiring rest in bed. Without the adrenal cortical response one could live quietly and in safety as long as the environment presented no physical or mental stresses, no important infections, injuries, or other challenges. But one could not go without food for long, nor carry out strenuous exercise; one could not *survive* severe injury or major surgery.[8]

Increases in corticosteroid production have been shown to occur in man during various mental as well as physical stresses.[9] For instance, in a university boat-race the coxswain had as great a response as the crew who were actually rowing.[10] On a 400 mile ski journey across Greenland lasting 20 days[11] the strongest member of the expedition had the lowest corticosteroid response except when he made a navigational error that forced a detour, and a university lecturer had a greater response during a busy autumn term than at any time during the 400 mile ski journey. Combat soldiers in the Korean war had as great a response on mere exposure to danger as when they were severely injured.[12]

Other mechanisms

The hypothalamus, via its action on the anterior pituitary gland, also increases the hormone output of the thyroid gland. This is what happens in stressed men or sheep, although there is a decrease in thyroid hormone output in animals such as the mouse or rabbit. There may also be an increase in the output of growth hormone from the anterior pituitary gland.[13]

There are probably other paths by which the stress response operates; for instance we do not understand how mental stress exacerbates certain skin diseases, arthritis, or asthma. There are also important changes in the brain giving rise to the conscious experience of stress. These include the mental changes of anxiety, the mental clarity experienced during examinations or interviews, and the apprehension and alertness that goes with a dry mouth and palpitations. Little is known about the mechanism of these mental changes, and important advances in the neuropharmacology and neuroanatomy of stress will doubtless occur within the near future.

When the stress is mental in origin it may seem inappropriate that the response is still largely physical, involving muscles, heart, and intestine. But these responses are ancient ones. They are present in all vertebrates and arose in evolution to deal with physical situations. In many animals and in man they also occur in response to social stresses. Their importance for survival has been such that they now fail to discriminate between physical, social, and symbolic stresses, between threats to the body and threats to self esteem or life situation. It is important to remember that the mental changes in the response are of great value, giving clearer thinking, resistance to mental fatigue, and resistance to the disrupting effect of conflicting emotional forces. Also, it is possible that these mental changes depend to a great extent on the circulating hormones that cause the bodily response. But it would seem inappropriate to make unconscious bodily preparation for a physical assault on one's friend during a heated armchair discussion, on the government after reading the morning newspaper, or on society itself in the face of a threat to one's life situation. The fact that modern mental stresses still call into action the apparently inappropriate response of muscle, heart, intestine, and skin is testimony, not only to the tremendous biological importance of the total response, but also to the very recent appearance, in evolutionary terms, of such mental stresses. Unless the inappropriate bodily response proved to be a severe selective disadvantage it might take several million years of genetic change in civilised man before the physical and mental components were dissociated. An alternative genetic change might enable man to monitor the mental stresses of modern life at the unconscious level and eliminate the irrelevant ones before they activated the stress response, in much the same way that irrelevant sounds are screened out before reaching consciousness.

Although the same general mechanisms are involved in the response to many different stress stimuli, there are differences. The relative importance of intestinal, skin, and heart changes, or of cortisol, thyroxin, epinephrine, and norepinephrine may depend on the stress, on the individual, and on the species.

DISEASES PRODUCED IN ASSOCIATION WITH THE STRESS RESPONSE

From the outset, it must be remembered that in most people most of the time the stress response, far from being disadvantageous or pathogenic, is vital and necessary. It seems to be only in the unlucky minority that there are pathological consequences. On the other hand, body and mind function as a whole, and for this reason alone mental stresses are likely to have some influence on the production and course of diseases.

This section includes almost all psychosomatic medicine, and constitutes part of the subject matter of psychiatry and psychoanalysis. It is a difficult field, and although pathological responses to stress are abundantly recognised in folklore, poetry, literature, and in the English language itself, this has not always been a respectable topic for serious scientific study. The evidence about stress has often been anecdotal, and many of the larger surveys have been retrospective, with other important influences uncontrolled or unconsidered. There have been those who were anxious to believe all, and those who have been reluctant to believe anything. One of the difficulties is that in a given patient with stress symptoms and disease it is often difficult to decide whether the stress produced the disease, or whether the stress symptoms arose in *response* to the disease, which had other causes. Nevertheless, the accumulated evidence is overwhelming that the mental stresses, the worries and anxieties of life, *can* play an important but ill-defined part in the production and in the exacerbation of mental and physical disease. Sometimes these stresses are hard to identify, if symbolic and subconsciously imagined, and we may have to turn to the psychiatrist or psychoanalyst to understand them.

It is worth pointing out that there is little evidence that disease can occur as a result of the response to *physical* stress. On the contrary, physical stresses are sometimes applied therapeutically so that the stress response will aid recovery from disease; and for most people the physical stress of exercise exerts a healthy influence on mind and body. Functionally, stress responses can be regarded as mobilising the body for action, or for resistance to physical insults. It is when there is no bodily action, or no physical insult to be resisted that there may be trouble. Hence disease in man may be associated with the stress responses to threatened insults (usually to the ego or life situation), and from the stress responses that lead to emotion without motion. Little is understood about the consequences of such imbalance between the mental and physical response, and this remains an important field for research.

The experimental evidence about the pathological effects of the stress response comes mainly from animals, but I shall not survey the growing mass of literature on this subject. In the animal studies, the stress stimuli have often but unavoidably been unnatural, as in the production of experimental neuroses in monkeys[14] or gastric ulcers in restrained rats.[15] Sometimes the stress stimuli have been more physiological, as in the death without significant injury after encounters between wild male rats or the reproductive failures seen in crowded rodents.[16] It may be noted that the effect of crowding, social interactions, and other stresses on reproductive performance seems to be of considerable significance in the regulation of populations of certain animals such as rats and mice.[17] Studies in man are few, but the occurrence of war amenorrhoea[18] and a few observations on the effects of stress in pregnant women[19] suggest that mental stress can influence reproductive performance in man. Nevertheless, human populations in fact continue to expand under crowded conditions, so perhaps the selective forces favouring the fertile have been stronger than any inhibiting effect of social stress on reproduction.

There is often a rise in blood pressure during mental stress. Can hypertension arise because of a sustained state of overreaction to the stresses of life? Unfortunately the question is a complex one, and in spite of some interesting and suggestive studies[20] it is not possible to reach clear conclusions. The exact role of stress in relation to the production of myocardial disease is also unsettled,[21] but its importance seems established. Recent work indicates that the major predisposing factors to myocardial disease are especially connected with education and background,[22] and it is possible that they originate in childhood or adolescence. The characteristic sufferer is described as having an extraordinary mental and physical alertness, a persistent desire for recognition and advancement, an intense sustained drive for achievement, and as being time stressed.[23] Once myocardial disease is present mental stress can precipitate heart attacks, as shown by the occurrence of 'anniversary infarctions'[24] on the anniversary of the day an important event, such as the death of a parent, occurred.

There is a vast and growing literature on the subject of stress and disease, and I will give two examples of the production of disease in man by mental stresses. The first is bereavement, or the so-called grief reaction. Since the one certain thing about life is that it will end in death, the stress of bereavement is one to which man will always be exposed. Close survivors, such as widows or widowers, show fairly well-defined mental symptoms and signs.[25] Also, for many months after the death they may suffer from a variety of psychosomatic diseases such as ulcerative colitis, arthritis, and skin diseases.[26] More dramatically, in Lindemann's[27] classic studies of the grief reaction after the Coconut Grove fire in Boston, a few bereaved patients experienced massive rectal bleeding at the funeral itself, representing an acute attack of ulcerative colitis. There is also an increased mortality in survivors,[28] the so-called 'broken heart' phenomenon. Parkes[29] goes so far as to suggest that grief may prove as important to psychopathology as inflammation is to pathology.

My second example is the phenomenon of illness clustering. In a careful study by Hinkle and Wolff,[30] the illnesses and significant life events in several thousand people were recorded over a period of 20 years. It was found that in a given individual illnesses of all kinds, not only peptic ulcers and neuroses but also bacterial infections and tumours, tended to occur in clusters, and there was a significant association of these illness clusters with stressful life situations. A similar study by Holmes and Rahe[31] is still in progress, and the same association has been found.

Where the stress response is associated with disease, it often seems to be the autonomic nervous system or thyroid gland, rather than the adrenal cortex, which is incriminated. But far too little is known to be able to eliminate the adrenal cortical response as a significant factor. For instance, there is still much to be learnt about the permissive and co-operative actions of corticosteroids, catecholamines, thyroxin, and other hormones.

PART PLAYED BY STRESS IN HUMAN EVOLUTION

Physical stress

Throughout human evolution men have been exposed to cold, hunger, infection, and trauma. The bones of palaeolithic man show fractures, chronic osteomyelitis, dental abscesses and so on. Palaeolithic woman, one might add, suffered the additional stresses of childbearing, and commonly died before the age of 30 years.[32] No-one lived to be very old.

These physical stresses have been of great importance in the genetic weeding out of those with poor bodily resistance to trauma, infection, cold, and hunger. The stress responses described above, present in our prehuman ancestors, have been selected for and have proved their worth during the evolution of man.

Many physical stresses have begun to disappear during the last 100 years, at least in a fortunate minority of mankind. The settled, more crowded life that followed the neolithic revolution led to a great increase in the amount of microbial and parasitic infections, but this, with the possible exception of respiratory and enteric virus infections, is now gradually being brought under control and is diminishing. On the other hand, malnutrition remains widespread, and injury is still common in accidents and in the wars that we continue to fight. Indeed, certain physical stresses in the form of drugs, pollution, and diseases of old age and civilisation, have appeared quite recently. On the whole, however, there is likely to be a continuing reduction in the total amount of physical stress to which men are exposed, although injury, for instance, will never disappear, and disease will always be a stress.

Mental stress

Under this heading I shall discuss stress in the form of threats to the ego or, using the broadest term, to the life situation. I call it stress because it tends to call into action the ancient and vital bodily mechanisms involved in the response to physical stress. This is a new type of stress that has *appeared* during human evolution, and the following is a speculative account of its development.

Early palaeolithic man lived mainly in the present. He may have known that he would die, that winter changed to spring. He could talk, even learn. But his thoughts were very much to do with his day-to-day existence. He had no clear picture of himself in relation to his past and his future. Also, he did not think of himself as separate from the landscape, the vegetation, and the animals. He did not see enough of other groups of men to worry whether he was better or worse off than they were.

With the comparatively sudden appearance of the neolithic way of life there were great changes. Men came to be concerned and preoccupied with the future—with future possibilities and threats. Perhaps this came inevitably with the coming of agriculture and the domestication of animals; with the sowing, reaping, feeding, and thought for the morrow. Man also began to see himself more in relation to, yet separate from, the environment. He began to build, plan, and worry about the future. Language, too, took a great leap forward, and became concerned with the future, with concepts rather than with immediate word pictures. In the beginning was the word. Was the word in fact *anxiety*?[33]

Modern man is obsessed with the future, its threats and its promises. Urbanisation and more complex social structures have led to an increase in the number of these threats and promises. He seeks, strives, and struggles. The difference between palaeolithic and modern man is not that palaeolithic man had no degree of self awareness or time awareness, but rather that modern man has these concepts in far greater intensity. He also has a more intense *response* to these concepts, a more acute experience of them. By analogy, one can say that the acute quality of pain, or indeed of pleasure, is in the intensity with which the primary sensation is handled in consciousness. Morphia, for instance, relieves pain not by affecting any change in the ability to sense pain, but by altering the conscious response to that sensation.[34] The conscious response to pain is changed in a comparable way under hypnosis or by the action of a placebo, so that pain ceases to be an acutely unbearable experience. In the same way, the glimmerings of self and future in primitive man became central components, acutely perceived in the mind of modern man, giving him a clearer picture of himself, his place in society, and the goals that can be struggled for and worried about.

This, then, is the *phylogenetic* development of self awareness and time awareness. One can also trace the *ontogenetic* development of these qualities in all men during infancy, childhood, and adolescence.[35]

Concern with the self and the future made it possible for there to be a new type of stress which could operate in men at the conscious or the unconscious level. Also, since it was a chronic concern with self and future, chronic stresses could be generated. These mental stresses have tended to become more severe in recent years, especially in competitive societies, and competition is a particularly human type of stress. Competitive societies, with fluid socio-economic groupings, provide ever-present opportunities for success or failure. It may be noted that experimental neuroses can be produced in dogs or monkeys when there are difficult choices between alternatives. Perhaps some people are especially vulnerable to the mental stress of choices between conflicting alternatives. If they are born into a society with rather rigid religious or social structure, with clearly prescribed status and paths of conduct, they might enjoy better mental health and feel less insecure than if they were born into a society where these things were ill-defined or changeable.

It seems unlikely that primitive man was free of mental stress. There are primitive societies which have not had the opportunity to be corrupted by civilisation in which fear of spirits and the supernatural, anxieties, phobias, and obsession are common.[36] In any case, in man as in some animals, a certain amount of mental stress is an inevitable accompaniment of family and social organisation.

But almost nothing is known about the mind of early man. Only a time machine can rescue palaeopsychology or palaeopsychiatry from the realm of inference and guesswork. Nevertheless there is a great temptation, reflected in most of the world's great myths and religions, to believe that man was originally happy and good. But something went wrong—a misplaced message, the fall, even the process of civilisation—and he lost this state of primordial grace. Man's past was good, his future may be good, but he is now miserably *en passage*. What is the evidence for primitive man? Skeletal remains suggest that on occasions he killed, sacrificed, mutilated, even ate his fellow men,[37] as many primitives do to this day. On the other hand he sometimes cared for old or injured people. Even the Neanderthals did this. There is the skeleton of a male in his late 40s with a withered amputated right arm, who was killed by a fall of rock in a cave; also that of a crippled toothless old male (la Chapelle aux Saints), who must have been fed by the others for years.[38] Certainly many modern primitives are gentle, non-aggressive, and appear to live in an idyllic state of harmony with nature.[39] But it may be incorrect to suppose that our palaeolithic ancestors were like this. Much current thinking about the effect of civilisation on the biology of man depends on the assumption that there have been no genetically determined mental changes in human populations since early palaeolithic times. While this may be true, with everything accounted for by the obvious and tremendous influence of environment and culture, it remains a possibility that there have been significant genetic changes, albeit small ones. The surviving living primitives seem often to have been pushed to the most inhospitable corners of the earth by more aggressive invaders. The timid Bushmen were replaced by the more belligerent Bantus. Such aggressiveness would have had a strong selective advantage, and we do not really know to what extent it may have been genetically rather than culturally determined.

Although primitive man was not free of mental stress, it is clear that there has been an extraordinary increase since palaeolithic times. This increase, I am suggesting, was an inevitable result of the increase in self awareness and time awareness that took place during human history. And the increase in human populations since the neolithic revolution, together with the appearance of more complex social structures and opportunities to possess objects, have further increased the opportunities for mental stress.

Is there any evidence that there have been changes in the ability to respond to stress during human history? Richter's[40] work on the differences between domesticated and wild rats poses the question as to whether primitive man had a more vigorous adrenal response to stress than does 'domesticated' modern man. The few observations which have been made show that primitive South American Indians,[41] South African Bantus,[42] and African Negroes[43] have *lower* resting adrenal hormone outputs than Europeans. The part played by diet and other cultural differences is not known. They could nevertheless have a greater *response* to stress, as suggested by the analogy of athletes who tend to have slow resting pulse rates. But if, as has been suggested above, the ability to respond to the physical stress of starvation, exertion, infection, and injury has continued to be important in recent human history, and if in addition human 'domestication' has led to the appearance of new stresses, it would seem unlikely there has been a significant weakening in the stress response of modern man. Such a weakening may well occur when the ability to resist stress ceases to be of selective advantage.

What is wrong with mental stress, and why do we not like it?

Does it lead to disease or a dulling of the enjoyment of life?

Disease. As discussed earlier, mental stress certainly seems to be associated with certain diseases, although in coronary disease, for instance, there is a complex web of causality in which diet, race, smoking, lack of exercise, and mental stress play parts which have not really been sorted out.[44]

Thinking of the harassed executive, with his high powered morning in the office, his glass of antacid milk and hypotensive tablets at lunch, his token exercise on the golf course at weekends, it is easy to say that he would be better off digging his garden, working one day a week, turning down his personal activity dial, as it were. But is this because he is not happy leading his type of life, or is it merely because we are worried about his health and longevity? If future advances in pharmacology could prevent his hypertension, heart disease, insomnia, or peptic ulceration, would everything be all right?

Dulling of enjoyment of life. Stress of the sort I am talking about goes with anxiety and worry. It is important to distinguish a healthy amount of anxiety and worry from a pathological amount. For instance, anxiety neuroses represent pathological reactions, in which irritability, fatigue, headaches, and insomnia are disruptive mental forces, shattering the patient's ability to deal with life situations and taking him to the psychiatrist.[45] It is a maladaptation to stress. On the other hand, there is a healthy amount of anxiety and stress, which we all experience. It goes with *caring* about things and about the future. There is nothing abnormal about being sad, worried, or anxious about an exam result, and it is natural to mourn a loved one, even if it is called a 'reactive depression'. This type of anxiety can also be creative, and has a great deal to do with incentives and with striving. If it were eliminated men would surely be reduced to the level of contented cows, chewing the cud, happily unconcerned with the fate of other cows, the future of pastures, or the future of cows. Perhaps the largest proportion of happy and contented people are to be found in institutions for the mentally defective. Even with schizophrenics, it is the *least* severely affected who have the peptic ulcers and hypertension.[46]

STRESS IN THE FUTURE

First, if we are to do anything worth doing we shall have to prevent the physical stresses of pollution and malnutrition, and at least reduce the physical stresses of infectious and parasitic diseases. After all, two-thirds of those in the world are still engaged in a day-to-day encounter with stresses of the most ancient and elemental sort.

We are all going to live in larger and larger urban areas. The landscape will be urban with rural oases set aside for food production and, one hopes, for recreation. It will be the great modern challenge to science and technology to construct these urban areas with freedom from certain mental stresses. But it can be done. We certainly *can* prevent noise, we can ensure personal privacy, and we can provide opportunities for diversity, recreation, and creation. When these things are taken care of, I doubt whether high density living itself will be stressful. On the contrary it may be richer, more varied, and even healthier than low density living. We evolved as groups of 20-30, but *can* live in larger groups in which our close acquaintances still only number 20-30.

If, then, we were to eliminate the stresses of noise, lack of privacy, and even the unnecessary stresses of the struggle for possessions and status, what would be left? First, there would always be the stress of the individual living in a community. During infancy and childhood, the developing self comes into conflict with a variety of parental and social restrictions, and it is from the encounter between these conflicting forces that the adult emerges. As long as men live in families and communities, this will continue to be a source of stress. Is there anything more? Just the stress of a full life. The stress of learning, loving, hating, grieving, caring passionately about anything, trying hard to do anything. These must stay as the irreducible minimum. It is in the striving as much as in the achievement itself that human satisfaction lies. Hans Selye, the great pioneer of the stress concept, dedicated his book *The Stress of Life* to 'those who are not afraid to enjoy the stress of a full life'.

Of course, it is only some people who respond to stress in a way that leads to hypertension, coronary disease, or anxiety neurosis—a way that can be regarded as maladaptive. Future advances in medical science will doubtless enable such people to live less dangerously. Many other people respond adaptively. The stress response enables them to deal successfully with situations, and they suffer no harmful effects in spite of the rises in blood pressure, cholesterol, cortisol, and catecholamines. If in the future we achieve the above-mentioned reduction in total stress load, leaving only the residual and desirable stresses of the full life, then I do not think that these stresses will lead to significant mental or physical disease.

There is a danger that in a world with about 10^{10} people, it will be inconvenient for us all to enjoy a full life. For the good of society we shall have renounced our right to use lethal weapons like the gun and the automobile, our right to make rubbish, noise, or children, our right to be alone in the wide open spaces of the world. There is a serious possibility that for the good of society, for the efficient running of a densely populated world, we may slowly be expected to give up our right to be different individuals, to do as many different things as possible, to have adventures, to seek and to strive. But these rights matter. They will always be a source of stress, but it is in the exercise of these rights that the human quality resides. A man without such stresses would be like a violin without strings.

Boredom, let it be remembered, can be a stress, and it is not inconceivable that intelligent man, one day in the future, will start to become obsessed with an aching awareness of his own nothingness. The best preventative is that we hold on to these basic rights. If ever there are moves to curtail them, or to produce men who are happy without them, we must resist. We must resist the human batteries as portrayed by E. M. Forster in his story 'The Day the Machine Stopped'. If necessary we shall have to take the world population back to 10^8. Better 10^8 human beings than 10^{10} human cows ('Homo insipiens'). It was Julian Huxley who asked the right question, 'What are people for?' All the important religions and philosophies have taught about the *quality* of life. The question of numbers had not arisen in those days, but if it had, the world's great teachers would have had no doubt which came first.

Summary

A stress stimulus disturbs homeostasis sufficiently for there to be a co-ordinated bodily response to deal with the situation. The response may involve physical action, as in running or fighting, or bodily resistance to cold, injury, or infection.

The stress responses arise in the central nervous system and are expressed through the hypothalamus, which in turn acts through the autonomic nervous system and the pituitary gland. The pituitary gland controls the secretion of adrenal cortical and other hormones. As a result of these responses there is a general mobilisation of the body machinery to deal with the stress. In evolution this has proved an appropriate and vital type of response which has been of immense adaptive value.

The stress response can take place when the need for action or for resistance to physical insults is foreseen in the central nervous system; the foreseeing of the need then in fact becomes the stress stimulus.

In man, and especially since the neolithic period, there has been a great increase in time awareness, giving greater opportunities for responses to foreseen stresses. An increase in self awareness increased the number and variety of stresses to the self. Also, advances in language and the ability to symbolise, meant that men could respond to imagined and symbolic stresses. Although such stresses were mental, they continued to induce the bodily responses appropriate for dealing with physical stresses. The possible mental stresses were further increased by the appearance of possessions and more complex societies, giving men responsibilities, status, and ownership.

Thus, while physical stresses have been important throughout human evolution, significant mental stresses have appeared since neolithic times.

Mental stress is associated with certain mental and physical diseases which are common in modern man, but there is often a complex web of causality in which stress and other factors play poorly understood parts. Some of the mental stresses afflicting modern man, such as noise, lack of privacy, inadequate recreation, can be reduced by appropriate urban planning. Others, such as the competitive struggle for status and possessions, could at least be reduced by socio-economic changes and it is likely that advances in medical science will render the stress response less pathogenic. If mental stresses are reduced, and if in addition there continues to be a fall in the physical stresses to which men are exposed, there will always remain the stress of a full life.

The stress of a full life is desirable and inescapable if men are to continue as men. If population pressures make it necessary to have unstressed (dehumanised) men, then quality must come before quantity and the world population must be allowed to fall.

References

[1] H. Selye, *The Stress of Life*, London, 1956, Longmans, Green.

[2] W. B. Cannon, *Bodily Changes in Pain, Hunger, Fear and Rage*, 2nd ed., New York, 1929, Appleton.

[3] S. Schachter and J. E. Singer, 'Cognitive, Social and Physiological Determinants of Emotional State', *Psychol.Rev.*, **69**, 1962, 379-99; F. Elmadjian, J. M. Hope, and E. T. Lamson, 'Excretion of Epinephrine and Norepinephrine under Stress', *Recent Prog.Horm.Res.*, **14**, 1958, 513-53.

[4] S. Wolf, *The Stomach*, New York, 1965, Oxford University Press.

[5] D. H. W. Kelly, 'Measurement of Anxiety by Forearm Blood Flow', *Br.J.-Psychiat.*, **112**, 1966, 789-98.

[6] H. Selye, *The Stress of Life*, London, 1956, Longmans, Green.

[7] R. W. Schayer, 'A Unified Theory of Glucocorticoid Action I', *Perspect. Biol. Med.*, **8**, 1964, 71-84; R. W. Schayer, 'A Unified Theory of Glucocorticoid Action II', *Perspect.Biol.Med.*, **10**, 1967, 409-18.

[8] L. P. le Quesne, 'The Response of the Adrenal Cortex to Surgical Stress', *Ciba Foundation Study Grps* No. 27, *The Human Adrenal Cortex*, 1967, 67-78.

[9] H. Persky *et al.*, 'Relation of Emotional Responses and Changes in Plasma Hydrocortisone Level After Stressful Interview', *A.M.A.Archs.Neurol.Psychiatry*, **79**, 1958, 434-47; J. W. Mason, 'Psychological Influences on the Pituitary-adrenal Cortical System', *Recent Prog.Horm.Res.*, **15**, 1959, 345-89; D. A. Hamburg, 'Plasma and Urinary Corticosteroid Levels in Naturally Occurring Psychological Stresses', *Res.Publs.Ass.Res.nerv.ment.Dis.*, **40**, 1962, 406-13.

[10] S. R. Hill *et al.*, 'Studies on Adrenocortical and Psychological Response to Stress in Man', *Archs.intern.Med.*, **97**, 1956, 269-98.

[11] H. W. Simpson, 'Field Studies of Human Stress in Polar Regions', *Brit.med.J.*, **1**, 1967, 530-3.

[12] J. Howard *et al.*, 'Studies of Adrenal Function in Combat and Wounded Soldiers', *Ann.Surg.*, **141**, 1955, 314-20.

[13] F. C. Greenwood and J. Landon, 'Growth Hormone Secretion in Response to Stress in Man', *Nature* (London), **210**, 1966, 540-1.

[14] J. N. Masserman, *Principles of Dynamic Psychiatry*, 2nd ed., Philadelphia, 1961, W. B. Saunders, 134.

[15] D. A. Brodie, 'Ulceration of the Stomach Produced by Restraint in Rats', *Gastroenterology*, **43**, 1962, 107-9.

[16] S. Barnett, 'Social Stress', *Viewpoints in Biology*, **3**, 1964, 170-218.

[17] J. J. Christian, J. A. Lloyd, and D. E. Davis. 'The Role of Endocrines in the Self-Regulation of Mammalian Populations', *Recent Prog.Horm.Res.*, **21**, 1965, 501-71; S. Barnett, 'Social Stress', *Viewpoints in Biology*, **3**, 1964, 170-218.

[18] F. E. Whitacre and B. Barrera, 'War Amenorrhoea', *J.Am.med.Ass.*, **124**, 1944, 399-403.

[19] D. H. Stott, 'Physical and Mental Handicaps Following a Disturbed Pregnancy', *Lancet*, **1**, 1957, 1006-12; B. B. Berle and C. T. Javert, 'Stress and Habitual Abortion', *Obstet.Gynec.N.Y.*, **3**, 1954, 298-306.

[20] N. A. Scotch, 'Sociocultural Factors in the Epidemiology of Zulu Hypertension', *Am.J.publ.Hlth*, **53**, 1963, 1205-13.

[21] W. Raab, 'Emotional and Sensory Stress Factors in Myocardial Pathology', *Am. Heart J.*, **72**, 1966, 538-64.

[22] L. M. Hinkle *et al.*, 'Occupation, Education, and Coronary Heart Disease', *Science*, N.Y., **161**, 1968, 238-46.

[23] M. Friedman and R. H. Rosenman, 'Association of Specific Overt Behaviour Pattern with Blood and Cardiovascular Findings', *J.Am.med.Ass.*, **169**, 1959, 1286-96.

[24] E. Weiss *et al.*, 'Emotional Factors in Coronary Occlusion', *Archs.intern.Med.*, **99**, 1957, 628-41.

[25] C. M. Parkes, 'Bereavement and Mental Illness', *Brit.J.med.Psychol.*, **38**, 1965, 1-26.

[26] C. M. Parkes, 'The Effects of Bereavement on Physical and Mental Health: A Study of the Medical Records of Widows', *Brit.med.J.*, **2**, 1964, 274-9.
[27] E. Lindemann, 'Symptomatology and Management of Acute Grief', *Am.J. Psychiat.*, **101**, 1944, 141-8.
[28] W. D. Rees and S. G. Lutkins, 'Mortality of Bereavement', *Brit.med.J.*, **2**, 1967, 13-16.
[29] C. M. Parkes, 'Bereavement and Mental Illness', *Brit.J.med.Psychol.*, **38**, 1965, 1-26.
[30] L. E. Hinkle and H. G. Wolff, 'Ecological Investigations of the Relationship between Illness, Life Experiences and the Social Environment', *Ann intern.Med.*, **49**, 1958, 1373-88.
[31] T. H. Holmes and R. H. Rahe, 'Life Crisis and Disease Onset', paper presented before Psychiat. Res. Soc. Palo Alto, Calif., 1966.
[32] H. V. Vallois, 'The Social Life of Early Man: The Evidence of Skeletons', in *Social Life of Early Man*, S. L. Washburn (ed.), Viking Fund Publications in Anthropology, 1961, Wennergren Foundation, 214-35.
[33] J. H. Masserman, in Proceedings of the 3rd World Congress of Psychiatry, R. A. Cleghorn (ed.), Montreal, 1961, McGill Univ. Press.
[34] H. K. Beecher, 'Non Specific Forces Surrounding Disease and the Treatment of Disease', *J.Am.med.Ass.*, **179**, 1962, 437-40.
[35] J. Piaget, *The Child's Conception of Physical Causality*, London, 1930, K. Paul.
[36] P. M. Yap, 'Mental Disease Peculiar to Certain Cultures: A Survey of Comparative Psychiatry', *J.ment.Sci.*, **97**, 1951, 313-27; E. D. Wittkower and H. Rin, 'Transcultural Psychiatry', *Archs.gen.Psychiat.*, **13**, 1965, 387-94.
[37] H. V. Vallois, 'The Social Life of Early Man: The Evidence of Skeletons', in *Social Life of Early Man*, S. L. Washburn (ed.), Viking Fund Publications in Anthropology, 1961, Wennergren Foundation, 214-35.
[38] C. S. Coon, *The History of Man*, London, Harmondsworth, 1967, Pelican, 62-3.
[39] G. P. Murdock, *Our Primitive Contemporaries*, New York, 1934, Macmillan.
[40] C. P. Richter, 'Domestication of the Norway Rat and Its Implication for the Study of Genetics in Man', *Am.J.hum.Genet.*, **4**, 1952, 273-85.
[41] H. W. Simpson, 'The Daily Adrenal Rhythm in Equatorial Amerindians', *J.Endocr.*, **32**, 1965, 179-85.
[42] W. M. Politzer and B. Tucker, 'Urinary 17-ketosteroid and 17-ketogenic steroid Excretion in South African Bantu', *Lancet*, **2**, 1958, 778-9.
[43] N. A. Barnicot and D. Wolffson, 'Daily Urinary 17-ketosteroid Output of African Negroes', *Lancet*, **1**, 1952, 893-5.
[44] C. M. Parkes, 'Bereavement and Mental Illness', *Brit.J.med.Psychol.*, **38**, 1965, 1-26; J. Stamler, 'Nutrition, Metabolism and Atherosclerosis', in *Controversy in Internal Medicine*, F. J. Ingelfinger *et al.* (eds.), Philadelphia, 1966, W. B. Saunders, 27-59.
[45] L. F. Chapman *et al.*, 'Highest Integrative Function in Man During Stress', *Res.Publs.Ass.Res.nerv.ment.Dis.*, **36**, 1958, 491-534.
[46] E. D. Luby, 'An Overview of Psychosomatic Disease', *Psychosomatics*, **4**, 1963, 1-8.

Comments

M. G. MCCALL Dr Mims has given us a stimulating and provocative paper. He has covered a broad canvas and I hope he will forgive me if I comment on matters of detail.

The effect of physical stress on endocrine activity and autonomic activity, with consequent effects on muscles, skin, blood vessels, cardiac output, digestion, haemostasis, and carbohydrate metabolism, is relatively well understood. It appears reasonable that in an evolutionary context, and in the light of the physiological changes evoked by stress, this reaction has continued because it enhances physical and mental capacity to deal with the stressful situation. That similar physiological changes are evoked by mental stimuli alone we can recognise in our day-to-day life, and it would certainly appear that civilised living conditions provide an increasing incidence of stress situations where physical action is inappropriate.

The crux of the matter is, however, the unequivocal demonstration in man that the stress reaction unaccompanied by appropriate physical activity is deleterious to health.

Kessel and Munro,[1] in reviewing the relevant literature in 1963 on psychosomatic disease for the WHO Committee on Mental Health, were forced to the following conclusion, and I quote:

> 'It has been the fate of the majority of studies to fail to show any significant relationship between the psychosomatic illnesses and personality, emotional factors, or psychologically stressful circumstances.'

They go on to note that this will not deter those committed to the banner of psychosomatics.

Again, Stammler,[2] reviewing the roles of stress, tension, high drive personality, and behaviour patterns in the genesis of coronary heart disease, notes that scientific understanding of the influence of these factors is far from clear. Stress is present in the lives of populations in economically underdeveloped countries, but they remain remarkably free of premature clinical atherosclerosis. In this context it is as well to remember that during World War II the incidence of clinical coronary heart disease fell substantially in a number of European countries, including England, Holland, and Scandinavia, at a time when stress of a particular type must have been very high.

It does appear, however, that stress, whether physical or mental, does exacerbate certain illnesses regardless of whether these illnesses are definitely organic or possibly psychosomatic. It also seems evident that in any urban population there tends to be a group of persons prone to both physical and mental ill-health.

Such socio-medical studies as are available, for example from Fanning,[3] suggest that rather than increasing stress being the sole basis of excessive morbidity in highly urbanised groups, boredom, particularly in women, may be responsible for a marked increase in morbidity and attendance at the doctor's surgery for a variety of complaints vaguely categorised as psychoneurotic. Furthermore, the organisation of life within high rise flat units tends to isolate people not only from social contacts but also from outdoor activities. Social isolation again contributes to psychoneurotic illness, while isolation from outdoor physical activities, particularly in children, is thought to be a factor in the flat dwellers' predisposition to excessive morbidity, for example from upper respiratory tract infections. Dr Freeman's comments on childhood freedom of movement in more primitive societies may have particular relevance to these observations.

On the basis of available information, including that reviewed by Dr Mims, I cannot think that we are yet in a position to do much more than speculate about the effects of the stress and isolation of modern urban life on physical and mental health of man.

Answers to the present gaps in our knowledge will demand a multi-disciplinary approach to the collection of new data, and the preservation of information about the entire range of activities and characteristics of communities which exist at present, as mentioned by Professor Stanley. Answers also lie in part in the re-organisation of data collected routinely at present so as to facilitate its use in research. The collection of uniform personal identification data on vital certificates and on health data would prepare the way for future record linkage studies. If, for example, linkage studies were possible between hospital morbidity data currently collected and death certificates there would be an end to the sort of difficulties Dr Wells has noted in the death certificate analyses presented by Dr Furnass. Not only would we know exactly what discrepancies exist in death certificates, but steps could be taken to rectify inaccuracies soon after the collection of the data.

There are few areas in the world where the whole range of civilisation, from populations of nomadic foodgatherers and hunters to high rise flat dwellers, is available for study. Australia is one of these rare natural laboratories where different cultural and ethnic groups exist side by side, and where, because of a rapid process of change and development, both secular and cross-sectional trends can be established.

In a situation where man is committed to increasing urbanisation but has it within his capacity to modify and develop the urban environment, it would seem imperative that, when there is suggestive evidence that an as yet undetermined degree of ill-health is a consequence of as yet poorly delineated factors in urbanisatilon as currently organised, further multidisciplinary studies on these problems be actively encouraged—and that in Australia we realise and exploit our potential for such research.

1 W. I. N. Kessel and A. Munro, WHO Expert Committee on Mental Health, Working Paper No. 28, 1963.

2 J. Stammler, Lectures on Preventive Cardiology, New York, 1967, Grune and Stratton.

3 D. M. Fanning, 'Families in Flats', *Brit.Med.J.*, 4, 1967, 382.

P. J. NESTEL Evaluation of stress is very difficult. It is a highly individual response which is variable and dependent on many factors such as the individual's personality, his intellectual and scholastic ability, the appropriateness of the stress, etc. Objective measurement of stress is a very inexact science. Many psychometric and physiological indices have been measured during stress tests and some of the following points are worth considering as examples of the difficulties which are encountered.

(1) There is no uniform response to different varieties of stress. Even the secretion of adrenalin and noradrenalin, which are often thought to exemplify stress most clearly, are not necessarily stimulated by the same kinds of stress. Adrenalin may be secreted predominantly in some people, especially following fear and anxiety, whereas noradrenalin may be secreted mainly in other people or in response to anger and aggression. Since one hormone may raise the heart rate and blood pressure and the other lower them, this point is of some importance. Furthermore, stresses may stimulate the pituitary adrenal axis and not the sympathetic nervous system and vice versa.

(2) We have studied the relationships between different physiological manifestations of stress such as the changes in blood pressure, heart rate, respiratory rate, sweating of the skin, and some biochemical changes such as the rise in plasma free fatty acids, the excretion of adrenalin, etc. Correlations do exist but they are of a low order ($r = 0{\cdot}3\text{-}0{\cdot}4$). Interestingly, estimates of personality such as neuroticism also show significant, though low-order, correlations with the physiological changes. Thus, the physiological changes bear some relationship to stress and to the subject's make-up, but the low correlation suggests inexact measurement and/or multiplicity of other factors.

(3) It is therefore not surprising that the response of an individual to stress is variable in some, but not in other respects. We have found, for example, that some manifestations of stress such as the rise in blood pressure and heart rate may be very similar to repeated exposures to stress, whereas the rise in adrenalin and plasma-free fatty acids may adapt, i.e. fail to rise on repetition.

(4) This merely emphasises the heterogeneous response to stress. It may be a reason why some people develop psychosomatic disease and others do not. However, the response within disease states is also far from homogeneous. Coronary artery disease and hypertension have been mentioned as possible stress-related disorders. We have found the response to stress to vary greatly among men who had survived a coronary occlusion. On the average their response was greater than in normal subjects. However, those men who subsequently suffered from chest pain showed greater sympathetic nervous responsiveness to stress in terms of blood pressure, heart rate, and noradrenalin secretion than those men who had no further chest pains after their occlusion. The increased response to stress in coronary disease may therefore have reflected the fear of pain or a predisposition towards developing pain rather than a causal relation to the disease itself. The difficulties in relating a disease to stress are also exemplified in hypertension. We measured the changes in blood and noradrenalin excretion

during stress. Though hypertensives on the average had significantly greater rises in blood pressure and noradrenalin, it was clear that this applied particularly to labile hypertensives who comprise only one of several varieties of hypertension.

These points demonstrate some of the problems in measuring and evaluating the response to stress especially as it applies to disease.

Discussion

RUBBO I was very interested in Dr Mims's separation of the different types of stresses and particularly in regard to his classification of physical stresses, all of which were really harmful stresses; stresses of starvation, stresses of disease and infection, and stresses of other injuries. It seems to me that in modern civilisation, we find a stratification of individuals in society so that, as you move towards the élite in the community, you find they are subjected more to the mental stresses of decision-making and less to the physical stresses of their environment; and they are subjected to other stresses, such as over-eating and so on. What I want to get at is this—surely there is one element of stress which no one seems to have mentioned and that is the question of healthful physical stress. In other words, are we as a community, a car-borne, lift-borne, and centrally-heated community, avoiding the challenge of physical stress? Is it true, for example, that in our present society those who are engaged in heavy hard work and not involved with a degree of mental stress are less subject to coronary heart disease and hypertension and other psychoneurotic manifestations of illness as compared to those who are subjected predominantly to mental stresses and less to the physical stresses?

PROVINS I have been very interested in the discussion so far which has centred around the long-term physiological effects of stress, but I should like to follow up the point made by the last speaker concerning the development of technology and the impact of associated changes (e.g. reduction of physical exercise) on our present way of life. To my mind, this raises the question of the effects on behaviour of the type and degree of physical stress which occurs in industry in particular. The degree of technological sophistication to which we have evolved today has almost inevitably brought with it noise stress, heat stress, and similar physical stresses which until recently appear not to have been given the systematic attention by psychologists that perhaps they deserve. However, in the past few years some fruitful experimental work has been undertaken on these conditions in relation to the efficiency of man in his working environment the results of which have, perhaps, theoretical implications relevant to the present discussion.[1] One of the findings seems to be that the effects of stresses such as heat, noise, and sleep loss interact, that is, their effects on the individual may be additive or they may subtract from one another.[2] These effects are shown in the performance of the individual, in the efficiency with which he carries out his daily tasks, and from the work of physiologists such as Magoun and others[3] on the reticular formation of the brain stem, a theoretical explanation has been postulated to account for the immediate or short-term effects of these stresses.[4] So far as I am aware no similar work has been carried out on the long-term or enduring effects of the action and interaction of such stresses on human performance (i.e. comparable to the investigations of Selye and Richter[5] at the neuro-endocrine level) although this is undoubtedly of relevance to the present pattern of living in industrialised societies.

1 D. E. Broadbent, 'Effects of Noise on Behaviour', in *Handbook of Noise Control*, C. M. Harris (ed.), New York, 1957, McGraw-Hill, Chapter 10.

2 D. E. Broadbent, 'Differences and Interactions Between Stresses', *Quart.J.exp. Psychol.*, **15**, 1963, 205-11; K. A. Provins, 'Environmental Heat, Body Temperature and Behaviour: An Hypothesis', *Aust.J.Psychol.*, **18**, 1966, 118-29.

3 G. Moruzzi and H. W. Magoun, 'Brain Stem Reticular Formation and Activation of the E.E.G.', *E.E.G.Clin.Neurophysiol.*, **1**, 1949, 455-73; D. B. Lindsley *et al.*, 'Behavioural and E.E.G. Changes Following Chronic Brain Stem Lesions in the Cat', *E.E.G.Clin.Neurophysiol.*, **2**, 1950, 483-98.

4 D. O. Hebb, 'Drives and the C.N.S. (Conceptual Nervous System)', *Psychol.Rev.*, **62**, 1955, 243-54.

5 H. Selye, 'The Physiology and Pathology of Exposure to Stress: a Treatise Based on the Concepts of the General Adaptation Syndrome and the Diseases of Adaptation', *Acta Inc.Med.Pub.*, 1950, Montreal; C. P. Richter, 'Neurological Basis of Responses to Stress', in Ciba Foundation Symposium on the *Neurological Basis of Behaviour*, G. E. W. Wolstenholme and C. N. O'Connor (eds.), London, 1958, Churchill.

WALKLEY My first comment relates to what Dr Freeman was saying about the different human behaviours induced by different plans of dwellings, when he drew attention to the lack of walls in a Samoan hut and compared this with the box-like, cage-like structure of the modern house, with its doors that are lockable. I think perhaps it is of some significance to point out that not only have dwellings in general over the last couple of generations become more freely planned, more openly planned, with fewer barriers in the form of walls and doors, but I think on the larger scale our towns and our cities are being more freely and more openly planned in the same way, with fewer confined spaces. I think this is perhaps relevant to the discussion. The other point I would appreciate making, Mr Chairman, is that as Director of the Australian Institute of Urban Studies, I was delighted to hear Dr McCall advocate further studies of higher density living and all its consequences—because this seems to me to be one of the outstanding problems of our cities of Australia. It is my job to try to determine some of the principal problems in many different fields besetting our cities, and this one—this matter of high density living as against low density—is one of those which crops up almost every day. Every city in Australia, even the small ones, have a problem great or large, relating to redevelopment; and redevelopment usually means replacing what exists with higher density housing as in Carlton, Melbourne and similar areas. I am sure that the planners of today—in particular those charged with the responsibility for redevelopment—would be very pleased to see a greater breadth and depth of study of the effects on human beings of higher density living, and it is in this field that I hope there may be further collaboration between the Institute I represent and some of the gentlemen here today.

FREEMAN A number of us seem to be agreed that Dr Mims's paper was stimulating; it stimulates me to speculate briefly on an evolutionary issue. Two of the points discussed by Dr Mims were (a) the nature of the pathways between the cerebral cortex and endocrine and related systems, and (b) the great antiquity, phylogenetically, of these latter systems. Now, if we consider the quite rapid evolution, in phylogenetic terms, of the human neocortex, it would appear that a phase was passed through during the upper palaeolithic when there was increasing input from the evolving neocortex into these systems. One of the principal discoveries of Freud was the mechanism of repression, and it is possible that the adaptive function of repression is the monitoring of input from the symbolising neocortex into the endocrine and other phylogenetically much older physiological systems. If this be so, a different conclusion may be drawn from that leapt to by some enthusiasts in the 1920s, who thought that the repressed state was 'unhealthy' and ought always to be 'got rid of', a conclusion associated with an attack on civilised values on the ground that 'civilisation spells repression'.[1] I think that most psychiatrists no longer take this point of view[2] and that it is now recognised that the uncontrolled loss of repression is prone to end in psychosis. I would disagree with Dr Mims on one minor point; we have not, I agree, a time machine, but this does not forever deny us any real understanding of the unique history of our species, for we do happen to have brains pre-adapted to the making of scientifically based inferences about the course of human evolution, and it behoves us to use them.

[1] S. D. Schmalhausen, *Why we Misbehave*, New York, 1928, Macaulay, 34.
[2] N. Cameron, *Personality Development and Psycho-pathology*, Boston, 1963, Houghton Mifflin, 238-9, 300-1.

HETZEL (*Chairman*) I do not think we have a psychiatrist here, and I think it is perhaps just as well in some ways because it would not be fair to have just one psychiatrist. I think we would need three or four of them to answer your question, Dr Freeman, because I do not think there would be a general consensus among them on the answer.

It is a very complicated question, and in a sense we envy Dr Mims his capacity to gather the whole landscape into one fine picture. But the qualifications introduced by Dr McCall and Dr Nestel arising out of experience of investigation of these matters in man are I think quite appropriate. The word 'stress' burst in upon us, of course, from physical science and was taken over and popularised by Hans Selye, and apparently as a result of his advocacy has been absorbed unchanged into all the European languages. He gave a lecture in Paris, I believe, in 1950 in which he used the word, and the word was accepted by the French which was perhaps not usual, with respect to

Professor Dubos! The French agreed it should be masculine! And I believe it occurs now in all the continental languages. It is a word that has its dangers, as I think many of us well know. It is of course a particular word referring to interaction of man with his environment, and it has been very loosely used, I think, by Selye himself, whose work was entirely confined to rats. Those of us concerned with stress in man get very uneasy when people talk too much about Selye and his particular work, which has, however, been a great stimulus in the field. The statement of the hypothesis linking stress and disease has to be very carefully made, in my opinion. We know that most diseases have multifactorial causation. A particular aspect of interaction with the environment is just one facet of the whole problem. We know that man, as we have been reminded here, has a very diverse genetic inheritance and particular patterns of physiological response of blood pressure, clotting time, gastric secretion, or whatever it may be, occur only in certain people and, with an appropriate trigger, lead to the occurrence of disease in certain susceptible subjects. The general hypothesis that there is a link between environment and disease I think we can accept, most of us, but there are a lot of details to be filled in. Dr Mims's paper seemed to me to provide a very useful sort of generalisation for an audience of this type, but I am afraid those of us in the field have many headaches before we have the precise knowledge we need. At the same time I would admit that we have been rather narrow in our approach to this problem. We have been thinking more in mechanistic than general biological or psychobiological terms, and I think we benefit from discussion of this sort. Now I think I will ask Dr Mims if he would like to say a few words just in closing.

MIMS I have no time to answer all of the points, so I shall say only a few words about what some people said.

I agree completely with Dr Freeman. There are important implications of the inability of the neocortex to discriminate between stress stimuli and to repress irrelevant ones rather than feed them all into the limbic system. In a very comparable situation the neocortex screens out irrelevant noises and so we do not consciously hear anything. It has been suggested that if the Middle Ages had lasted ten thousand years there might (from the evolutionary point of view) have been time for the neocortex to reach some adjustment to civilised life and screen out irrelevant stress stimuli.

We certainly need to quantify the study of stress. I had to oversimplify, present a broad canvas which anybody could understand, and my paper was not, of course, a detailed technical description or discussion of stress and its mechanisms. We need to quantify and we need to find out a lot more. On the other hand, the fact that quantification is difficult does not mean that stress cannot be studied and that no conclusions can be drawn from such studies even now. We are not in the enviable situation of the zoologists who have a little animal called the tree shrew; when stressed, the hairs on its tail stand on end, and you can measure the proportion of each twelve-hour period that they do this.[1] This is beautifully quantitative, and can be correlated with the stress results in terms of retarded growth, reproductive performance and so on. Unfortunately, it is very hard for man, who has no hairs on his tail!

[1] H. Autrum and D. von Holst, 'Social Stress in the Tree Shrew (*Tupaia glis*) and its Effect on Growth, Body Weight and Reproduction', *Z.Vergl.Physiol.*, **58**, 1968, 347-55.

The Impact of Civilisation on the Biology of Man

Cultural Adaptation to Biological Maladjustment

The capacity to adapt to changing conditions is an essential property of living matter, and the fate of species in evolution and of individual organisms in their lifetimes is determined largely by the degree of effectiveness of their adaptive processes.* The mechanisms by which adaptation takes place are multiple and varied, but each falls fairly readily into one of three broad categories. Before embarking upon the main topic of this paper, which concerns a form of adaptation of exceptional importance to the human species, it is relevant to summarise the characteristics of these three classes of adaptive processes.

Evolutionary adaptation

This kind of adaptation consists of the modification of the genetic constitution of populations through natural selection so that they become better fitted, in the Darwinian sense, to the prevailing conditions of life. When lasting changes occur in the environment of a species, new selection pressures are introduced and the species eventually either becomes genetically adapted to the new circumstances or else it becomes extinct, the latter being the more usual outcome.[1]

Thus the sum total of the inherited characteristics of an individual animal—all the features of its anatomy, its physiology, its innate behaviour—are the product of this kind of adaptation acting on the ancestral populations of the species during its evolutionary history.

In the case of *Homo sapiens*, the genetically-determined characteristics which distinguish the species from the other primates, including his high degree of dexterity, his ability to communicate through learned symbolic speech, his exceptional capacity for learning, and his unusual intelligence are the products of the evolutionary pressures which were operating throughout the miocene and pliocene periods, but especially during the two million years of the pleistocene period.[2]

* According to the *Shorter Oxford Dictionary*, 'adaptation' means 'the process of modifying so as to suit new conditions'. In general, the usage of this word in biological, medical, and psychological literature comes within this definition, although there is a tendency nowadays in some circles for the term to be used for *any* response of the organism to environmental influences. This widening of the meaning of the term is confusing and unnecessary, and in this paper we restrict its use to those changes which take place in populations or in individual organisms that render them better suited to the environment and better able to cope with the conditions of life.

Innate or genetically-coded adaptation

Under this heading we include all those inbuilt responses which occur spontaneously in the individual animal in the face of environmental change and which render the organism better able to cope in the new conditions. The responses may be physiological, behavioural or, in the case of the young animal, they may affect growth and development.*

Some of these genetically-coded reactions occur rapidly as, for instance, in the case of the chemical changes which occur in the blood following sudden fright or anger and which render the individual better equipped to meet the physical demands of the new situation. Others, such as the healing of wounds, are slower in developing.

Repeated stimulation of sense organs results in a temporary reduction in their sensitivity, and in some physiological textbooks the term 'adaptation' is used exclusively for this phenomenon. Thus the temporary impairment of hearing which occurs in man when exposed to high levels of noise in factories and elsewhere can be regarded as adaptive in the noisy situation itself, although it is clearly maladaptive when the affected individual tries to understand ordinary conversation in quieter surroundings.[3] Even the permanent deafness which follows frequent and prolonged exposure to high levels of noise might be considered adaptive in that it may have a beneficial effect on the individual's performance in the noisy environment.

Instinctive patterns of behaviour which occur spontaneously in animals in response to certain environmental stimuli, such as the flight reaction following the appearance on the scene of a potential predator, fall into this second category of adaptive processes.

Genetically-coded mechanisms of adaptation are, of course, the products of evolutionary adaptation. This fact is important in the present context, because it tells us that such adaptive processes can exist for coping only with environmental predicaments of a kind that have been within the fairly frequent experience of the species during its evolutionary history. An unnatural environmental situation may well trigger off one or more of the innate 'adaptive' mechanisms, but these are likely to be ineffectual or even harmful in the strange circumstances.[4]

Adaptation through learning and conditioning

Increasing in importance as we ascend the evolutionary scale is a kind of adaptive behaviour which is not genetically-coded, and which depends on the processes of learning and on the previous experience of the individual. Learning achieves its greatest significance in *Homo sapiens,* in which species it is responsible for by far the greater part of the individual's behaviour.

* It is worth emphasising that all the responses in the second category are 'genetic' in the sense that the sequence of events in each case is determined by the animal's hereditary constitution. For this reason, it is preferable to avoid the usual term 'genetic adaptation' for the first category of adaptive responses, since this usage tends to give the misleading impression that the other kinds of adaptation are independent of genetic control. The term 'physiological adaptation' which is often used for the second category of adaptive reactions is unsatisfactory because it disregards behavioural adaptive responses.

'Cultural adaptation', the subject of this paper, belongs to this third category of adaptive processes. It occurs almost exclusively in man and depends not only on man's ability to learn from experience, but also on the capacity of human society to accumulate knowledge gathered by its members and to pass this knowledge on to other individuals and to subsequent generations. While the processes of cultural adaptation are not genetically coded, they are, of course, dependent on man's genetically-determined potential to learn and to communicate through the use of symbolic language.

Although it is well recognised that man, as compared with other species, has available to him this additional means of adapting to environmental conditions, the processes of cultural adaptation receive extraordinarily little serious attention from natural scientists, social scientists, or students of the humanities. The purpose of the present paper is to stimulate some discussion in this neglected area. It is concerned primarily with cultural manoeuvres aimed at overcoming forms of biological maladjustment which arise in the human organism as a consequence of changing conditions of life associated with the processes of civilisation.

THE IMPORTANCE OF CULTURAL ADAPTATION IN CIVILISATION

Through natural selection, species develop biological characteristics which render them extremely well-suited to the conditions to which they are exposed as they evolve. Should the conditions of life deviate in any way from those to which the species is thus fitted through evolution, signs of maladjustment are almost inevitable. The more drastic the changes, or the more numerous they are, the greater the level of maladjustment that can be anticipated and in the long run the survival of the species will depend on the effectiveness of the mechanisms of adaptation to the new conditions.

It requires little imagination to appreciate that numerous extensive changes have occurred in the conditions of life of man during the past few thousand years. These changes involve every aspect of his biology, including his diet, his sleep patterns, the quality of the air he breathes, the number and closeness of daily contacts with other members of his species, his sexual environment, and so on. Signs of biological maladjustment have inevitably appeared in response to these changes and it is pertinent to inquire into the relative importance of the roles of the three categories of adaptation in assisting the species to survive under the new conditions.

With regard to the role of evolutionary adaptation, the significant fact is that only three or four hundred generations have passed since the time when all mankind led a nomadic life based on hunting, food-gathering and the use of tools and weapons fashioned by hand out of raw materials provided by the natural environment. In as small a number of generations as this, appreciable genetic change could have occurred through natural selection only in the presence of fairly strong selection pressures. In general, however, the new environmental conditions imposed on the human organism since the neolithic development have not, under the protective conditions of civilisation, exerted pressures of this order. An exception to this statement has been the introduction of powerful new selection pressures resulting from the greatly increased incidence of infectious disease associated with the conditions of civilisation, and there is good evidence that certain microbial agents have influenced the genetic constitution of human populations with respect to genes associated with resistance or susceptibility to the diseases in question.[5]

On the whole, however, too few generations have passed since the first beginnings of civilisation for appreciable genetic change to have occurred in the human species in response to the new conditions. Recently, of course, spectacular environmental changes have occurred within the living memory of many people, and any possibility of genetic adaptation to such recent events is out of the question. Another point of relevance is that, although cities have been in existence for some two hundred generations, until recently only a small proportion of the world's population has lived in them. Most of the genes of modern urban populations have taken the rural route to the city.

It would be unreasonable, of course, to maintain that the frequencies of different genes in the human population as a whole are exactly the same today as they were during the late palaeolithic period. But the changes that have occurred have been due mainly to historical events acting, for example, through migration, genetic drift, and sexual selection. They have not, by and large, been the result of natural selection operating in response to new environmental pressures associated with civilised living.

The pace of cultural change has thus outstripped by far the rate of possible evolutionary adaptation. There has been no time for natural selection to produce a genetically new breed of men better adapted physiologically and mentally to the conditions of life which now prevail in modern Western cities. We must look, therefore, to the other two categories of biological adaptation for the responses which have enabled the species to survive the rapidly changing conditions.

The genetically-coded adaptive mechanisms have obviously been brought into play continuously since the neolithic development, as indeed they had been during the palaeolithic period. On the other hand, these mechanisms can hardly be credited with man's relatively successful adaptation to the new and special conditions associated with civilisation, since many of the changes have been of a kind, or of a degree, that could not have come within the normal range of experience of his evolutionary ancestors.

The fact that neither evolutionary adaptation nor adaptation through genetically-coded responses can possibly explain the survival and multiplication of the human species in the new conditions helps us to appreciate the supreme importance for civilised man of the third category of adaptive mechanisms—those which depend on learning and in particular on cultural processes.[6]

The recognition by society of the existence of a disorder which is considered undesirable because it is painful, because it causes death or for some other reason, is usually followed by remedial counter-measures. It is this process that concerns us in this paper; it can be represented diagrammatically in its simplest form, as follows:

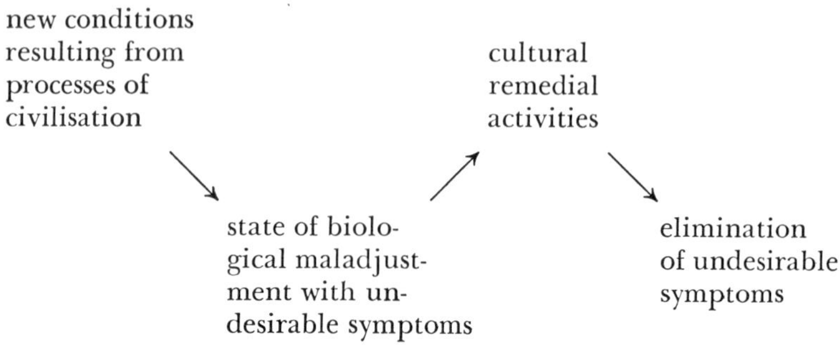

This general pattern can be illustrated by two straightforward examples. Firstly, economic factors and technological advances have resulted in the introduction of substantial amounts of refined carbohydrates into the diet of modern man. This change in the conditions of life, imposed on the human organism by cultural developments, gives rise to various signs of biological maladjustment of which one of the most clear-cut is the development of cavities in the teeth,[7] a disorder which for several reasons society recognises as undesirable. Economically, the most important cultural response to this particular form of maladjustment is the establishment of a professional body of men and women who are highly skilled in the art of filling the cavities. The Australian economy supports over 3,000 of these persons, as well as their technical assistants, secretaries, and the manufacturers of the various tools of their trade.[8]

The second model concerns the population situation. During the late palaeolithic era, the total population of *Homo sapiens* in the world is thought not to have exceeded ten million,[9] and it is reasonable on biological grounds to assume that most of the population was, for most of the time, well nourished. We do not know in the case of man, as we do of some other species, what were the innate physiological or behavioural mechanisms that controlled population growth under the conditions of nature, but there are sound reasons for suspecting that, as in other mammalian species, inherited (non-cultural) mechanisms existed that tended to regulate the reproductive rate in relation to the availability of food supplies. Clearly the artificial conditions of civilisation have interfered with the operation of these regulatory mechanisms and, as we know only too well, population growth can now continue in spite of inadequate nutrition. The over-crowding, continued malnutrition, and human misery that result can reasonably be regarded as manifestations of a state of biological maladjustment on the community level. The deliberate cultural adaptive responses to this form of maladjustment include all manner of procedures ranging from infanticide to various methods of contraception. Incidentally, authorities who speak against the use of contraceptive procedures on the grounds that they are unnatural show a striking inconsistency of argument in that they express no similar objection to the unnatural conditions that permitted the population crisis to develop in the first place.

In passing, it should be noted that biological maladjustment in human society frequently has important social and economic repercussions other than those which are adaptive or intended to be adaptive. For example, it has been suggested by some historians that the Black Death, which was due to the particular biological conditions prevailing in Europe during the fifteenth century, played an essential role in precipitating the end of the feudal system in England.[10] There are countless examples of important social, political, and economic consequences of biologically-induced disorders.[11]

TWO TYPES OF CULTURAL ADAPTATION

It is important at the outset to recognise that there are two distinct kinds of cultural adaptation to biological maladjustment, differing in principle with regard both to their modes of action and to their long-term effectiveness. The first of these we call 'corrective' and the other 'antidotal' cultural adaptation.*

Corrective adaptive processes are those which aim to reverse the underlying change which is ultimately responsible for the state of biological maladjustment. Antidotal adaptation occurs when an attempt to overcome a disorder is directed at a symptom of the disorder or at an intermediate and subsidiary cause but not at the unsatisfactory biological conditions which gave rise to the state of maladjustment in the first place.

Corrective cultural adaptation is seen, for instance, in moves to restore essential nutrients to the diet when symptoms of malnutrition occur, in the introduction of quarantine procedures to combat certain contagious diseases, and in the ritual of performing physical exercises before breakfast to 'keep fit'. In the case of dental caries mentioned above, corrective cultural adaptation would consist of replacement of refined carbohydrates in the diet with 'natural' unrefined foodstuffs. The profession of dentistry and the addition of fluoride to the water supply are both antidotal adaptive processes.†

Antidotal cultural responses include the administration of antibiotics to overcome infectious disease, and the extensive use of tranquillisers, sleeping pills, and the other psychotropic drugs. Indeed the pharmaceutical industry, which represents one of the most impressive examples of cultural adaptation to biological maladjustment in modern society, depends for its existence almost entirely on the antidotal

* 'Correct' = 'to make or set right'; 'antidote' = 'whatever tends to counteract evil that something else might produce', *Webster's Collegiate Dictionary*, 5th ed., Springfield, 1943, Merriam.

† The distinction between 'corrective' and 'antidotal' is not the same as the distinction between prophylactic and therapeutic. Corrective adaptation may, as in the case of treatment of scurvy with ascorbic acid, be therapeutic, and antidotal adaptation may be prophylactic as in the case of vaccination against infectious disease. In general, however, corrective adaptive processes tend to be prophylactic and antidotal to be therapeutic.

approach to biological maladjustment. In economic terms, the cost of this cultural adaptive response to biological maladjustment amounts to over $A200 million per year in Australia* and more than $US6,000 million in the United States of America.[12]

Corrective adaptation is more likely to be successful than antidotal adaptation in the long run, although it is not in all cases a feasible proposition—either because social or economic factors prohibit it or because the underlying biological cause of the disorder is unknown. Antidotal cultural responses tend to be tail-chasing in character in that they very often create as many new difficulties as they solve. The complications which have followed the widespread use of antibiotics are well known. While these substances have undoubtedly been very effective in reducing the mortality due to various bacterial diseases in some areas of the world, it is well to recall that public health measures, such as the improvement of housing, sanitation, and nutrition, and the introduction of quarantine procedures—all good examples of corrective adaptation—have played a vastly greater role in reducing the incidence of infectious disease in Western society than have the antibiotics, and they have not given rise to undesirable repercussions.

There are other serious disadvantages to the antidotal approach. Consider, for example, the situation with regard to the psychotropic drugs, notably the tranquillisers and anti-depressants, which today form the basis of a $US300 million industry in the USA.[13] If prevailing environmental conditions induce a high frequency of behavioural disorders that can be easily alleviated by these drugs, the disagreeable likelihood exists that their widespread use will permit the continued progressive deterioration of the environment until eventually the dose of the drugs has to be doubled, and later trebled and so on.

The activities of the medical profession represent another important aspect of cultural adaptation to biological maladjustment. A number of surveys have been carried out in recent years on the nature of the illnesses which cause people to visit their doctors[14] and an examination of the findings shows clearly that the majority of patients are suffering from 'diseases of civilisation', in the sense that they are complaining of disorders which would have been exceedingly rare or non-existent in palaeolithic society. The work of medical practitioners in Western countries thus consists mainly of countering signs of maladjustment which result from the prevailing biological conditions. Most of the advice they give amounts essentially to advocation of various forms of antidotal adaptation, and it is not surprising therefore to find that a further substantial part of the effort of the medical profession is spent in countering the undesirable biological consequences of procedures used in the treatment of disease.[15]

* During the year 1965/6 the cost of 'benefit' prescriptions was $A94,630,000 and of prescriptions to hospitals and miscellaneous authorities was $A14,634,000 (information from *J. Commonwealth Dept Health*, **16**, No. 3, Sept. 1966). The cost of 'non-benefit' prescriptions was $A24,000,000 (information from Pharmaceutical Manufacturers Information Bureau). The total is $A133,264,000. This figure does not include medicines sold 'over the counter' and without prescriptions.

THE IMPORTANCE OF KNOWLEDGE OF THE BIOLOGICAL PROCESSES UNDERLYING DISEASE

Needless to say, society is in a much better position to develop rational means of dealing with manifestations of biological maladjustment if it possesses knowledge of the true nature of the disorder in question, in terms of *all* the contributing factors. Nevertheless, lack of such information does not exclude the possibility of successful adaptation.

In the complete absence of all knowledge concerning the causation of a sign of maladjustment, a period of fumbling counteractivity ensues, based partly on incorrect interpretations of the situation and partly on trial-and-error tactics. Man is usually convinced that there does indeed exist a cause behind his unwanted afflictions, and when in the past his knowledge of the material world has not provided him with a ready explanation, he has, not unnaturally, tended to invoke divine or spiritual intervention.[16] Thus, cultural attempts at adaptation to maladjustments resulting from conditions imposed on man by civilisation have not infrequently involved intensified religious activities of one sort or another. In England in 1666, after the Great Plague and the Fire of London, there was a widespread feeling that these calamities were divine punishment for neglect of religious matters by the community, and a Bill was introduced in Parliament designed to curb atheism in the country. Although the Bill was eventually dropped, Parliament continued to be influenced in its actions by this general attitude. Thus, among the somewhat bizarre cultural responses aimed at preventing a recurrence of the state of biological maladjustment manifest in the plague, we can list the refusal of the authorities to allow the English philosopher, Hobbes, whose views were frequently at variance with those of the Church, to publish any of his writings on religion or morality.[17]

The trial-and-error approach has often been successful in the discovery of effective procedures for dealing with forms of biological maladjustment in the absence of knowledge concerning their true nature. There are many stories of the efficacy of various concoctions prepared from herbs by primitive peoples for the treatment of divers ailments. One of the most successful of these was cinchona bark, which was discovered by the indigenous inhabitants of Peru to be very effective in the treatment of malaria. It was brought to Europe by the Jesuits in 1632 and the revolution it brought about in the history of European medicine was described by Ramazzini as comparable only to the introduction of gun-powder in the art of war.[18] The active principle of cinchona bark is quinine, which persisted as the specific treatment for malaria until very recently.

It was partly the effectiveness of cinchona in the treatment of malaria that led Edward Stone in the mid-eighteenth century to investigate the medicinal properties of extract of willow (*Salix*) bark, which has a bitter taste similar to that of cinchona bark. He was also motivated by the ancient doctrine that the cures of maladies often occurred naturally in the same regions as their causes. In 1763 Stone presented to the Royal Society his paper entitled 'An Account of the Success of the Bark of the Willow in the Cure of Agues'.[19] The Hottentots in southern Africa had independently discovered, presumably through trial-and-error, the therapeutic value of 'willow tea'.[20] One of the constituents of extract of willow bark is salicin, which was later found to be very effective in pure form in the treatment of certain febrile and rheumatic conditions and also as a mild analgesic.[21] Today the famous derivative of salicin, acetyl-salicylic acid or aspirin, is probably the most widely used single chemical compound in man's efforts to overcome biological maladjustment; in the USA alone some thirty million pounds of this substance are consumed annually.[22]

The undoubted effectiveness of aspirin satisfactorily explains its widespread and continued use as an antidotal remedy. It is more difficult to understand the long lasting popularity of that greatly prized therapeutic and prophylactic agent of the Middle Ages, known as Venice treacle, which contained the flesh of vipers as one of the most important of its fifty or more various ingredients. It originated in the second century B.C. and was still in use in India in 1835.[23] Indeed it is said to be sold by at least one pharmacist in Venice at the present day. The mixture was used not only for the therapy of all kinds of sickness, but also for preventive purposes, and was highly recommended for protection against plague in the seventeenth century.[24] Horn of unicorn was another ancient medicament which retained its popularity for many hundreds of years; it was still recommended in the 1677 edition of the London Pharmacopoeia for such forms of biological maladjustment as plague, measles, and smallpox.[25]

Successful adaptive procedures arising from the empirical approach are usually, although not always, of the antidotal variety. An example of the trial-and-error method leading to successful corrective adaptation is seen in the discovery in the seventeenth and eighteenth centuries that scurvy could be successfully treated with, and also prevented by, the consumption of fresh fruit and vegetables. The scurvy story illustrates another variable which influences cultural adaptation–the lag period between the discovery of an effective means of preventing a certain maladjustment and its implementation. It was in 1753 that James Lind published his famous *Treatise on Scurvy*[26] in which he set out the convincing evidence that scurvy could be prevented by 'greens or fresh vegetables, with ripe fruits', advocating their inclusion in the diet of seamen among whom scurvy was a perpetual problem; but it was not until over forty years later that the Lords of the Admiralty adopted Lind's recommendations.[27]

In general, however, attempts at cultural adaptation are unlikely to be successful when society remains in complete ignorance of the nature of the disorder that it is trying to overcome. But as time goes on and new facts come to light, a stage of partial understanding is achieved, and it has been on the basis of such partial understanding that most of the more effective cases of cultural adaptation have developed. One of the best examples is the public health movement of the last century in Britain. This was aimed primarily at endemic and epidemic diseases such as typhoid, dysentery, and cholera which were extremely prevalent in certain sections of the community at the time.

We believe now that the high incidence of these diseases was due to the effects of the prevailing ecological conditions on the interplay between the human organism and certain microbes, permitting the evolution, spread, and perpetuation in human populations of highly pathogenic strains of bacteria, viruses, and protozoa.

The men responsible for the social reforms which led to the general cleansing of the cities and the improvement of community health rightly noted that the diseases in question appeared to be associated with certain kinds of living conditions which were described so vividly in Chadwick's *Report on the Sanitary Conditions of the Labouring Populations of Great Britain*, published in 1842.[28] However, they were wrong in their belief that the diseases were caused by 'odours' given off by decomposing organic matter.

John Snow, the great epidemiologist who so clearly linked the outbreak of cholera in London in 1854 with the contaminated water supply from the Broad Street pump, did in fact conclude that this disease was due to a self-multiplying living agent which lived in the human body and which was spread mainly by means of the excreta of infected persons.[29] The authorities did not accept this view and, although they enforced Snow's recommendations for improved sanitation, stated that 'under the specific influence which determines an epidemic period, fecalised drinking water and fecalised air may breed and convey the poison'.[30]

Thus, while the public health measures of the latter half of the nineteenth century were successful in countering biological maladjustment to the extent that the incidence of typhus, cholera, and tuberculosis dropped spectacularly, this success was achieved, at least at the beginning of the period, without knowledge of the role of micro-organisms in the disease processes.

At the present time there are in Western society a number of well recognised and serious manifestations of biological maladjustment, the nature of which is partially understood and which nevertheless show little or no signs of disappearing. An example is cardiovascular disease, which accounts for about one-third of the deaths in the adult male population in Australia, North America, and the United Kingdom today. The evidence is convincing that this complex of disorders is in some way associated with the increased standard of living and the 'Western way of life',[31] although the precise factors responsible for the maladjustment remain a matter of uncertainty and controversy.

Another sign of maladjustment that is incompletely understood at the present time is criminal behaviour, including delinquency in the younger age groups. Some readers may question the reference to this topic in a paper on biological maladjustment, assuming that because it is a problem usually studied by social scientists and social psychologists it has nothing to do with biology. It is true that in the literature on the subject, antisocial activity is seldom discussed in terms of biological maladjustment. For example, in 1955 and in 1960 international congresses were held under the auspices of UNESCO on the subject of 'The Prevention of Crime and the Treatment of Offenders' and a large number of experts discussed the possible causes and prevention of juvenile delinquency; not one of them raised the possibility that biological factors might be playing a role in the aetiology of this kind of behaviour.[32] Meanwhile it is certain that the rate of crime of almost all kinds is increasing, that the cause of the increase has not been established and that the methods used for dealing with the situation have so far failed to be effective.[33] In these circumstances there seems some justification for the view that biologically-determined maladjustment may contribute to the increase in criminal behaviour, and that neglect of the biological approach to the problem may be contributing to the lack of progress towards its understanding and solution.

Obviously, the more complete our understanding of the nature and causes of a physiological or behavioural disorder, the better are the chances of successful cultural adaptation; but even complete understanding is not alone sufficient to guarantee that appropriate counteraction will be taken. Tuberculosis is a form of maladjustment which is now well understood. Certain kinds of living conditions are known to favour the spread and development of the disease, and the properties of the specific microbial agent that multiplies in the tissues have been studied in much detail. While the prevalence of tuberculosis in European cities is at present very low, in certain other areas of the world the disease continues to be a major cause for concern. Thus, in the African quarter of Johannesburg, the incidence of pulmonary tuberculosis is eighteen times higher than it is among the European population.[34] Such facts as these serve to emphasise that, while the understanding of the true nature of forms of biological maladjustment is of paramount importance in cultural adaptation, it is nevertheless only part of the story. Economic, social, and political factors determine on what scale the social reforms or medical treatment that are indicated are carried out, and at what rate.[35]

Perhaps one of the most spectacular examples of cultural adaptation in recent years has been the virtual elimination of malaria from many areas of the world where it was extremely prevalent only ten or twenty years ago. The economic and social consequences of this adaptive process have been far-reaching.[36] The anti-malaria campaign has been based on a fairly complete understanding of the nature of this particular form of maladjustment, and it has consisted primarily of an attack with the new insecticides on the mosquito vector of the disease. This approach, let it be noted, is not a corrective one, and it is thus not without undesirable side-effects, such as the development of insecticide-resistant strains of mosquitoes, ecological disturbances due to effects on other insect species, and the possibility of direct toxicity of increasing concentrations of insecticide in the environment on other forms of life, including man. Corrective adaptation in the case of malaria would consist of draining the man-made marshes in which the mosquitoes breed or simply the avoidance by human beings of the swampy areas.

THE COMMUNITY, THE INDIVIDUAL AND THE ROLE OF EDUCATION

Most of the adaptive responses mentioned above involve steps taken by the community, or by sections of the community, to counteract various manifestations of biological maladjustment. The state has played an extremely important role in the process, by introducing laws, for example, designed to ensure that certain standards of housing and sanitation are met, or by distributing funds to encourage scientific research on socially important problems.

The counter measures introduced by state authorities are often, but not always, examples of corrective adaptation, as in the case of most of the public health measures instituted by governments. It is also seen in their approach to the problems arising out of abuse of the habit-forming drugs. The ready accessibility of this group of pleasure-promoting and biologically very potent substances is a relatively new consequence of the processes of civilisation, and it is one that certainly gives rise to states of biological maladjustment. The actions of the state in bringing in legislation to restrict the sale and distribution of habit-forming drugs can thus be regarded as efforts to restore a situation that prevailed prior to the advent of civilisation. That it is the *availability* of these drugs, rather than lack of knowledge of their dangers, that is the more important factor in the aetiology of this form of biological maladjustment, is indicated by the fact that the incidence of drug dependence is higher among doctors and nurses than among the rest of the population, despite the fact that the members of these professions must surely, because of their educational background, be more aware of the undesirable effects of these agents than the rest of the community.[37]

Nevertheless, education is in general an extremely important aspect of successful cultural adaptation. When new knowledge is acquired which throws light on the causation of a disorder and indicates suitable curative or preventive measures, the relevant information must be transmitted from specialist scientific circles to other interested parties, be they doctors, administrators, or plain citizens. Very often cultural adaptation must involve positive action on the part of the individual himself. The state cannot take responsibility for seeing that every person consumes his necessary supply of vitamins every day; its role in such instances is to make sure that everyone in the community is properly informed, through education, of his nutritional needs and of the ways by which they may be met. One of the features of the present century has been the acceptance by governments of their responsibility in disseminating knowledge about nutrition.[38] Other sections of the community may also take on an educational role of their own accord as in the case of many societies and citizen groups formed to combat various forms of ill-health.

In some countries governments have recognised an educational responsibility in relation to the habit of smoking; they have not only instituted educational programs in schools and arranged for the production of films presenting information on the undesirable effects of smoking, but have also taken steps to curb the propaganda of the cigarette companies, which are seeking by all the subtle psychological means at their disposal to increase the number of cigarettes smoked in the community.[39] Other governments do not accept this responsibility and openly encourage the development of the industry in question. Once again we see that an understanding of the underlying causes of a form of biological maladjustment is by no means the only requirement which must be satisfied before society takes suitable steps to overcome it.

Thus, in considering the role of education in relation to cultural adaptation, it is important to appreciate that the education of the average individual in modern Western countries consists, as far as his biological needs are concerned, of two quite distinct facets. On the one hand, he receives a certain amount of relevant information and wise counselling at school, or through the family doctor, an occasional sensible magazine article or radio program; on the other hand he is exposed to a vast bombardment of extremely persuasive propaganda generated by commercial interests which are not always particularly concerned about his health and well-being.

As already mentioned, commercial forces are also very important in relation to the processes by which modern man adapts to the various kinds of maladjustment. Indeed the pharmaceutical industry depends for its existence on biological maladjustment and its cost to the community is a fair measure in economic terms of the level of biological maladjustment in the population, although we should add to this, of course, the cost of the medical and dental services. This total amounts to about $US30,000 million in the USA and $A1,000 million in Australia annually.[40] These figures do not include, of course, the economic loss due to absenteeism, and it could be argued that at least part of the cost of the police force, and perhaps even of the defence forces, should be added to the bill for biological maladjustment.

However, the role of commerce, in establishing biological maladjustment and in providing antidotal and often tail-chasing adaptive measures to overcome it, is too large a subject to attempt to tackle here. Let it be stated, however, that on occasion industry can be persuaded to spend money on the development of corrective methods of adaptation, as in the case of the relatively successful efforts to control certain forms of air pollution in many cities in Europe and the USA and in the efforts now being made by car manufacturers in America to develop a vehicle that gives off far smaller quantities of undesirable fumes than do existing models.

Returning to the question of cultural adaptation on the level of the individual, it is worth drawing attention to what is one of the most significant of the biological consequences of civilisation—the fact that 'doing what comes naturally' is no longer equivalent to 'healthy living'. Under the new conditions the individual must exercise a considerable degree of self-discipline in order to avoid ill-health. This is well illustrated by the situation which has developed with regard to physical work. Cultural developments over the centuries, and especially in the last few years, have tended steadily to decrease the amount of physical work performed each day by the average city dweller. While the evidence that this change gives rise to biological maladjustment is not absolutely conclusive, it is nevertheless very suggestive, and anecdotal evidence strongly supports the view that some regular physical exercise is necessary for optimal health in most people. Thus the city dweller may, for the sake of his health, discipline himself to run around the block every morning before breakfast, although his natural inclination may be to stay in bed as long as possible. Palaeolithic man did not have this problem, for he got his physical exercise without any act of self-discipline when his hunger drive finally overcame his natural lethargy.

THE INCREASING IMPORTANCE OF MILD FORMS OF BIOLOGICAL MALADJUSTMENT

We have seen that, as civilisation proceeds, it is bound to introduce changes in the biological conditions of the human organism, and that consequently various forms of biological maladjustment can be expected to become manifest in the population. The disorders which result often become the targets of adaptive processes set in motion by the community.

It is immediately apparent, however, that some forms of maladjustment tend to persist in society for much longer periods than others. Those most likely to persist are mainly of two kinds: *mild forms* which, although recognised as undesirable, are not sufficiently disagreeable, or are not of sufficient economic significance to warrant the marshalling of social forces to counter them. Among virus infections, for example, mumps, influenza, and numerous minor diseases of the upper respiratory and the alimentary tracts persist in Western society while the much more dangerous smallpox does not; and *insidious and chronic forms* of maladjustment which develop so slowly in an individual as he gets older, or in the population, generation by generation, that they are not even recognised as abnormal (which indeed they are not, under the new conditions).

Thus, modern urban man does not seem to regard fairly frequent headaches, a certain degree of malaise, mild disturbances of the digestive tract, frequent colds and other virus infections, and a considerable degree of irritability and aggressiveness in the home as out of the ordinary; and yet the evidence that a high prevalence of these states has always been the lot of man is completely lacking, and indeed there are good biological reasons for doubting such an assumption. That forms of maladjustment can pervade human society unrecognised is illustrated by the fact that many medical textbooks describe an increase in blood pressure with increasing age as a normal occurrence—a natural and inevitable feature of growing old. And yet it has been discovered in recent years that in some primitive societies the blood pressures of old and young men are similar, and there is now a strong argument for the view that increase in blood pressure with increasing age should be listed among the consequences of environmental influences associated with the civilised way of life.[41]

We can postulate, therefore, a sort of 'natural selection' taking place among the different kinds of biological disorders that arise in society as a consequence of the changing conditions of life. Forms of maladjustment which interfere drastically with the business of living or which threaten early death to a fairly high proportion of the community will be the most likely to engender an effective cultural adaptive response, and these will tend to be eliminated early. Milder kinds of maladjustment which are not recognised as disorders because of their gradual onset are likely to persist in society for much longer periods, and we can therefore anticipate that, as civilisation proceeds, there will occur a steady accumulation of different varieties of mild, insidious, and chronic forms of physiological or behavioural disturbances.

Unfortunately, the hypothesis that there is a progressive increase in the number of mild afflictions due to different causes associated with the progress of civilisation is a difficult one to prove; the data that would be necessary to do so are of a kind that are very hard to collect in a satisfactory way. Linked to this is the added difficulty that criteria for assessing quantitatively and objectively many of the milder manifestations of biological maladjustment do not exist. The best we can say at present is that the data that are available are at least compatible with the hypothesis and there are a number of arguments in its favour.[42]

It is possible for modern society to carry a substantial load of biological maladjustment because of the protection and support civilisation affords the individual who is physically or mentally unwell. A degree of ill-health that would have interfered severely with an individual's chances of survival and of successfully rearing healthy children in the Old Stone Age may have no such effect on Darwinian fitness under the new conditions of civilisation. In other words, it is now quite possible for the human species to multiply at an unprecedented rate, while at the same time experiencing a general state of ill-health that would never have permitted reproductive success under palaeolithic conditions. Thus, while it is reasonable to state that one of the effects of civilisation on the biology of man has been to increase his numbers some three hundredfold, it is unscientific to conclude that the species is therefore well adapted, physiologically and psychologically, to the new circumstances. The current population increase tells us little of man's general state of health and well-being in the modern world, as compared with palaeolithic times when his fertility was of a much lower order.

The possibility that the advance of civilisation is associated with the progressive accumulation in the population of different forms of mild and chronic biological maladjustment has social implications which deserve serious attention. Although the disorders themselves may be mild, their very ubiquity may well give rise to social repercussions far beyond the direct effect they have on the affected individual himself. Perhaps the least important of these, in terms of human well-being, is the economic cost to the community of widespread absenteeism and lowered productivity. Of far greater significance are the possible consequences for society of the effects of almost universal malaise on mental processes and personality factors; for there can be few, if any, periods in the history of civilised man when clarity of thought, high quality in decision-making, and mutual tolerance and understanding have been more important for the welfare of humanity than they are in the modern world. Although surprisingly little work has been reported on the psychological effects of mild or chronic ill-health, there is nevertheless sufficient evidence for us to be able to state categorically that physiological disturbances can have deleterious effects on intellectual performance and the quality of inter-personal relations.[43]

CULTURAL ADAPTATION IN THE FUTURE

Several factors point to the need for society to adopt new approaches to the problem of biological maladjustment. In the first place, the processes of cultural adaptation that we have seen operating up to the present have been singularly unsuccessful in eliminating the causes of minor disorders, although as these accumulate they may collectively become at least as serious in their overall effect as the more spectacular afflictions of civilised man. Moreover, the cultural response to milder forms of ill-health at present consists mainly of the discovery and widespread distribution of biologically potent chemicals effective in alleviating symptoms or sometimes in curing the disease in question. These compounds themselves frequently give rise to further disturbances of one sort or another and it is clear that this kind of antidotal approach cannot go on indefinitely without some very unpleasant consequences.

Another factor calling for a new approach is the greatly accelerated *rate* at which man's environment and conditions of life are changing as a result of the current technological revolution. The changes in human biology that have taken place since the neolithic revolution began three hundred generations ago are very numerous, and many of them have been introduced within the last few decades. They include such items as changes in the level of physical work; changes in the level of sexual stimulation; the wearing of clothes for purposes other than keeping warm; changes in eating and sleeping patterns; and changes in the number of contacts with infectious agents, with other animals, and with members of the same species. A particularly significant set of changes has been collectively referred to as the increasing 'chemicalisation' of the environment.[44] More and more biologically strange chemical substances are gaining access to the tissues of the human body, mainly but not exclusively via the alimentary canal and the lungs. The example of DDT is often quoted, and it has been estimated that some twenty tons of this substance alone are distributed in the livers of the citizens of the USA.[45] But there are many others and it has been said that, at a conservative estimate, 500 new chemical substances known to be toxic are added to our resources every year[46] and are liable to find their way in small quantities into our bodies. It has recently been pointed out that the intake by the human organism of metallic cations, such as chromium, aluminium, copper, and nickel, has increased enormously as a consequence of technological developments.[47]

The present situation demands that, instead of waiting as in the past for signs of maladjustment to reach such a pitch that the community demands some remedial action, we should in the future attempt firstly, to identify as many as possible of the changes that have taken place and are taking place in the biological conditions of life of the human organism; secondly, we must try to determine what, if any, are the signs of maladjustment resulting from each of the changes and what are their social implications; thirdly, we must ask what are the best means, in view of all the knowledge available to us and in terms of the well-being of mankind, of countering forms of biological maladjustment considered to be undesirable, so that in the long run the cultural advantages of civilisation are not heavily outweighed by its biological disadvantages.

The precise form which investigations on these problems should take is too large an issue to attempt to tackle in this paper. Laboratory experiments on animals will no doubt continue to provide illuminating and pertinent information. In addition, multidisciplinary, multifactorial studies of human populations are likely to play an increasingly important role, especially cross-cultural surveys involving comparisons between modern Western societies and other human groups following simpler patterns of life. A number of very interesting cross-cultural investigations of this kind are now under way, and we hope that the parameters selected for measurement include those which will prove most crucial to our understanding of the biological responses of the human organism to the conditions of modern civilisation; for time is running out, and insecticides, automobiles, and aspirins are spreading at a spectacular rate into even the most remote parts of the world.

Perhaps the most urgent need of all, however, is not for new information, whether it emanates from the laboratories or mass surveys, but rather for a great deal more intellectual effort aimed at ensuring that the wisest possible use is made of knowledge that has already been acquired for the benefit of mankind.

Summary

The development of civilisation has brought about drastic changes in the conditions of life of the human organism. As would be anticipated these changes have given rise to a whole range of signs of biological maladjustment. The survival of the species, and its success in overcoming the effects of such maladjustment under these new conditions, cannot be explained in terms of natural mechanisms of adaptation; it depends on man's capacity to adapt through cultural processes. Basically, there are two distinct kinds of cultural adaptation: *corrective* adaptation, which aims to reverse the underlying changes in conditions that are ultimately responsible for the state of biological maladjustment (e.g. the reintroduction of fresh fruit and vegetables to prevent scurvy and most public health measures against bacterial infection); and *antidotal* adaptation, which occurs when an attempt to overcome a disorder is directed at a symptom of the disorder or at an intermediate or subsidiary cause, but not at the unsatisfactory biological conditions which gave rise to the state of maladjustment in the first place (e.g. aspirins, antibiotics, tranquillisers, vaccination, etc.).

It is postulated that since cultural adaptive processes are usually directed against only the more alarming disorders, there has been occurring with the advance of civilisation a steady accumulation of different mild and chronic forms of biological maladjustment. While singly such relatively mild disorders may not appear particularly serious, collectively their effects on society may be quite far-reaching.

The greatly accelerated rate of environmental change associated with recent technological achievements calls for a new approach to the problem of the biological needs of the human organism in modern society.

References

1 J. Maynard Smith, *The Theory of Evolution*, London, 1958, Penguin Books, 249.

2 J. Buettner-Janusch, *Origins of Man*, New York, 1966, Wiley; G. A. Harrison *et al.*, *Human Biology*, Oxford, 1964, Clarendon Press; L. S. B. Leakey, 'An Early Miocene Member of Hominidae', *Nature* (London), 1967, **213**, 155-63.

3 A. Bell, 'Noise: An Occupational Hazard and Public Nuisance', *Publ.Hlth Pap.WHO.* No. 30, 1966; L. L. Beranek, 'Noise', *Scient.Am.*, **215**, 1966, 66-76.

4 R. Dubos, *Man Adapting*, New York, 1965, Yale University Press, 29-31; D. A. Hamburg, 'The Relevance of Recent Evolutionary Changes to Human Stress Biology', in S. L. Washburn (ed.), *Social Life of Early Man*, Chicago, 1961, Aldine, 278-88.

5 R. Dubos, *Man Adapting*, New York, 1965, Yale University Press, 185-90; G. A. Harrison *et al.*, *Human Biology*, Oxford, 1964, Clarendon Press, 86-9.

6 R. Dubos, *Man Adapting*, New York, 1965, Yale University Press, 12-13.

7 D. R. Brothwell, 'Teeth in Earlier Human Populations', *Proc.Nutr.Soc.*, **18**, 1959, 59-65; R. A. Ferguson, 'Some Observations on Diet and Dental Disease', *J.Am.dent.Ass.*, **22**, 1935, 392; J. L. Hardwick, 'The Incidence and Distribution of Caries throughout the Ages in Relation to the Englishman's diet', *Br.dent.J.*, **108**, 1960, 9-17; T. W. B. Osborn and J. N Noriskin, 'The Relation between Diet and Caries in the South African Bantu', *J.dent.Res.*, **16**, 1937, 431.

8 Preliminary figures from the 1966 Census, supplied by the Commonwealth Bureau of Census and Statistics, Canberra.

9 E. S. Deevey, 'The Human Population', *Scient.Am.*, **203**, 1960, 194-204.

10 G. M. Trevelyan, *Illustrated English Social History*, London, 1949, Longmans Green.

11 R. Dubos, *Mirage of Health*, New York, 1959, Harper, Chapter 7; W. H. S. Jones, *Malaria: A Neglected Factor in the History of Greece and Rome*, Cambridge, 1907, Bowes and Bowes; E. P. Prentice, *Hunger and History*, New York, 1939, Harper; H. Zinsser, *Rats, Lice and History*, London, 1935, Routledge.

12 M. L. Gross, *The Doctors*, New York, 1966, Random House, 484.

13 M. L. Gross, *The Doctors*, New York, 1966, Random House, 488.

14 See 'Trends in the Study of Morbidity and Mortality', *Publ.Hlth Pap.WHO.* No. 27, 1965, Geneva, for references.

15 D. P. Barr, 'Hazards of Modern Diagnosis and Therapy—The Price We Pay', *J.Am.med.Ass.*, **159**, 1955, 1452-6; D. M. Spain, *The Complications of Modern Medical Practices*, New York, 1963, Grune and Stratton.

[16] C. Coury, 'The Basic Principles of Medicine in the Primitive Mind', *Med.Hist.*, II, 1967, 111-27.

[17] B. Russel, *A History of Western Philosophy*, London, 1946, George Allen and Unwin.

[18] F. H. Garrison, *An Introduction to the History of Medicine*, Philadelphia, 1929, Saunders, 290.

[19] E. Stone, 'An Account of the Success of the Bark of the Willow in the Cure of the Agues', *Phil.Trans.R.Soc.*, **53**, 1763, 195-200.

[20] F. Ensor, 'The Willow as a Remedy for Acute Rheumatism', *Lancet*, **1**, 1876, 910.

[21] M. Gross and L. A. Greenberg, *The Salicylates: A Critical Bibliographical Review*, New Haven, 1948, Hillhouse Press, 5.

[22] H. O. J. Collier, 'Aspirin', *Scient.Am.*, **209**, 1963, 96-108.

[23] G. W. Corner, 'Methridatium and Theriac, the Most Famous Remedies of Old Medicine', *Johns Hopkins Hosp.Bull.*, 1915, 222-5.

[24] D. Defoe, *Journal of the Plague Year*, 1722, republished, Oxford, 1928, Blackwell, 292.

[25] W. Brockbank, 'Sovereign Remedies: A Critical Depreciation of the 17th Century London Pharmacopoeia', *Med.Hist.*, **8**, 1964, 1-13.

[26] J. Lind, (1753) *Treatise on Scurvy*, 1953 edition, C. P. Stewart and D. Guthrie (eds.), Edinburgh, University Press.

[27] C. P. Stewart in his introduction to *Treatise on Scurvy*, Edinburgh, 1953, University Press.

[28] E. Chadwick (1842) *Report on the Sanitary Condition of the Labouring Population of Great Britain*, reprinted with introduction by M. W. Flinn, Edinburgh, 1965, University Press.

[29] J. Snow, (1855) 'The Mode of Communication of Cholera' in *Snow on Cholera*, being a reprint of two papers, New York, 1936, The Commonwealth Fund.

[30] W. H. Frost in his introduction to *Snow on Cholera* (See reference 29).

[31] F. H. Epstein, 'The Epidemiology of Coronary Heart Disease, A Review', *J.chron.Dis.*, **18**, 1965, 735-74.

[32] First United Nations Congress on the Prevention of Crime and Treatment of Offenders, New York, 1956; Second United Nations Congress on the Prevention of Crime and Treatment of Offenders, New York, 1961.

[33] W. A. Lunden, *Statistics on Delinquents and Delinquency*, Illinois, 1964, Thomas.

[34] Report on the health of Johannesburg in 1965, Medical Officer of Health Report (1965).

[35] K. Evang, 'Political, National and Traditional Limitations to Health Control', in *Health of Mankind*, G. Wolstenholme and M. O'Connor (eds.), Ciba Foundation Symposium, London, 1967, Churchill.

[36] R. Barlow, 'The Economic Effects of Malaria Eradication', *Amer.Econ.Rev.*, **57**, 1967, 130-57; P. Newman, 'Malaria Eradication and Population Growth with Special Reference to Ceylon and British Guiana', Bureau of Public Health Economics, Research Series No. 10, School of Public Health, University of Michigan, Ann Arbor, 1965.

[37] M. M. Glatt, 'Problems Common to Alcoholism and Drug Dependence', *WHO Chron.*, No. 21, 1967, 293-303.

[38] C. Ware, K. M. Panikker, and J. M. Romein, *History of Mankind*, UNESCO Vol. 6, 1966, 457-8.

[39] 'Cigarette Advertisements and Television', *Med.J.Aust.*, **1**, 1967, 397-8.

[40] B. Abel-Smith, 'An International Study of Health Expenditure', *Publ.Hlth Pap.WHO.*, No. 32, 1968, 56; B. McFarlane, *Economic Policy in Australia: The Case for Reform*, F. W. Cheshire, Melbourne, 1968.

[41] A. A. Abie, 'Physical Changes in Australian Aborigines Consequent upon European Contact', *Oceania*, **31**, 1960, 140-4; B. Kaniner and W. P. W. Lutz, 'Blood Pressure in Bushmen of the Kalahari Desert', *Circulation*, **22**, 1958/60, 289-95; K. Maddocks, 'Possible Absence of Essential Hypertension in Two Complete Pacific Island Populations', *Lancet*, **2**, 1961, 396-9; L. J. J. Nye, 'Blood Pressure in the Australian Aboriginal', *Med.J.Aust.*, **2**, 1937, 1000; H. M. Whyte, 'Body Build and Blood

Pressure of Men in Australia and New Guinea', *Aust.J.exp.Biol.med.Sci.*, **41**, 1963, 395-404.

42 See, for example, R. F. L. Logan, 'The Need for Assessment of Sickness and Levels of Health', in *Trends in the Study of Morbidity and Mortality*, *Publ.Hlth Pap.WHO.*, No. 27, 1965, 9; I. H. Pearse and L. H. Crocker, *The Peckham Experiment*, New Haven, 1947, Yale University Press.

43 J. R. Clayer and M. N. Dumbrill, 'Diabetes and Mental Illness', *Med.J.Aust.*, **1**, 1967, 901-3; R. J. M. Collis, 'Physical Health and Psychiatric Disorder in Nigeria', *Trans.Am.phil.Soc.*, **56**, Pt 4, 1966, 1-43; K. Dalton, *The Premenstrual Syndrome*, London, 1964, Heinemann Medical Books; H. Geist, *The Psychological Aspects of Diabetes*, Springfield, 1964, C. C. Thomas; S. V. Kasl and S. Cobbs, 'Some Psychological Factors Associated with Illness Behaviour and Selected Illnesses', *J.chron.Dis.*, **17**, 1964, 325-45; A. Keys *et al. The Biology of Human Starvation*, Vol. II, Minneapolis, 1950, University of Minnesota, 767-918; F. N. Pitts and J. N. McClure, 'Lactate Metabolism in Anxiety Neurosis', *New Engl.J.Med.*, **277**, 1967, 1329-36.

44 T. G. Randolph, 'The Toxic Environment', in *Interactions of Man and his Environment*, B. H. Jennings and J. E. Murphy (eds.), New York, 1966, Plenum Press.

45 'Restoring the Quality of our Environment', Report of the Environmental Pollution Panel of President's Science Advisory Committee, The White House, 1965, 4.

46 D. H. K. Lee, 'Environmental Health and Ecology' in *Man—His Environment and Health*, Supplement to *Am.J.publ.Hlth*, No. 54, 1964, 7-10.

47 B. Bubela, personal communication, 1968.

Comments

EARLE HACKETT I have no fault to find with Dr Boyden's classification of our cultural adaptations—'antidotal' and 'corrective'. Dichotomies can mislead, but he allows grey areas in between. And I agree that corrective adaptations are likely to be biologically suitable because they adjust our environment to the *status quo ante*, to which we are presumably genetically adapted already. This is a modern scientific rational justification of Rousseau's irrational doctrine of the Noble Savage.

But antidotal adaptations should not be discounted just because they are so. Several speakers yesterday, notably Dr Wells, pointed out that the non-communicable diseases of civilisation are still fairly rare. Only small minorities, in a biological view, get serious cardiovascular disease, or lung cancer, or diabetes, or depression or neurosis, or go in for juvenile crime. We must not confuse the clinical or medical concern for the individual with the broader and much less personal view of the biologist, which is the view set down for this symposium. Now, susceptibility to the non-communicable diseases seems to have an inherited basis of some kind. I think this is just another indication of the large amount of genetic de-specialisation or polymorphism in our species. Polymorphism is apparent in all studies of human blood groups, plasma proteins and histocompatability genes, and seems likely to be present at many of our gene loci. The reason, I think, is obvious enough. Look at what has happened to us humans since we left those pleistocene lake shores in Africa. We have moved in social groups of fifty or more through so many environments in only a thousand generations (the number a sub-cultured bacterium can go through in a fortnight)—from cool to warm to hot, from dry to wet, from desert to forest, from seaboard to mountain top, from eating flesh to bread to blubber to sugar, and so on. And only those human communities that had a wide reserve of gene patterns could survive such changes and that meant the widely polymorphic communities, and this is our main biological adaptation to the process of *getting* civilised. And today our stock, exploding almost unchecked, is packed with genetic reserves of polymorphisms which can quite rapidly rearrange in heterozygous balances to suit changing conditions. Now it also follows that a minority of homozygous (or less frequently heterozygous) patterns in individuals are ill-preadapted (if I may use that phrase) to a particular civilised environment and are therefore showing, in medical terms, 'disease'. The very same people may well be beautifully preadapted to some other civilised environment yet to come. This is the biological argument in favour of what Dr Boyden calls the antidotal approach. But it is a form of human conservation and, as he says, it is called medicine.

A corollary to this is that not only should we study diseased people but we should also study those who are exceptionally non-diseased. When a pathologist does a necropsy on someone civilised who is found to have no atherosclerosis at all, he should blood group him extensively, freeze some of his still living cells in liquid nitrogen, try to arrange a full clinical examination of the blood relatives, and so on, and record the findings. A dentist who finds a city dweller with no caries at all in his teeth should do the same. Elderly people who retain exceptional athletic or intellectual ability should also be carefully examined. We need biological, chemical, and physical specifications of these abnormal people.

A related point is that not only are there, as Dr Boyden says, isolated primitive peoples, dwindling in numbers, who should be studied by the methods of comparative pathology, but there are dissident exclusive groups within our civilised community who adopt special biological policies. There are complete vegetarians. There are the Jehovah's Witnesses who forbid blood transfusion. As a doctor, the latter exasperate me, but as a biologist I can see that for virological or possibly ecological reasons it is just possible that we may need some human stock one day which has kept itself utterly free of blood injections for many generations. Social freedoms and biological dissent can be important for maintaining biological diversity.

Now if I may turn to the propagandist element that Dr Boyden touches on. He is prepared—wrongly I think—to give credit for the nineteenth century sanitary revolution to a 'partial understanding' (his phrase) of biological principles. I think this is wrong. He should give the credit to his irrational predecessor, Jean Jacques Rousseau—already mentioned this morning by Dr Freeman—who wrote of the superiority of the Noble Savage in his two Discourses to the Academy of Dijon in the mid-eighteenth century. The idea spread rapidly among top people in Western Europe. In England 'Capability' Brown laid out idyllic 'wild' country estates, and within fifty years the Lake poets had appeared and a movement had begun which aimed to let sweet smelling country air and sunshine into the slums and stews of the industrial cities. Smells, which meant faeces, were disposed of down drains. The sanitary revolution was under way, before Chadwick, long before Pasteur. Dr Boyden says that 'these important and effective measures were instituted when the understanding of the true nature of the diseases was still very incomplete'. He is wrong. They were instituted when there was *no* understanding, in our biological scientific sense. Rousseau's doctrine, reinforced in England by a puritanical belief that cleanliness is next to godliness, was surely just a random and quixotic turn taken by Western European culture. When the germ theory of communicable disease came along fifty years later the popular mind grasped it fairly quickly, but again, let me suggest, only because the popular mind was culturally preconditioned by a couple of thousand years of irrational thinking that disease was caused by evil spirits that got in from outside.

We must remember, as scientists considering health propaganda—and this is a matter for pessimism—that it *is* the irrational colourful ideas that have always captured and inspired the total human imagination and led to action. The Pope, whatever we think of his knowledge of sexual relationships, is backed in his authority by the utterly unrelated irrational image of an unmarried god crucified unfairly. The result is that an authority like the Pope can promote important social biological policies, be they right or wrong, far more effectively than any one here or than any World Health Organization using rational argument.

The trouble is that every single person in the community has his *own* ideas of human biology, and this hinders the general taking up of applied human biological science in the community. In contrast, the physics of flight or television, or the chemistry of dyes or plastics has only to pass the checkpoint of a handful of scientists and technologists before being applied within a decade or so. But the whole dead weight of humanity is made up of amateur applied biologists directing the living organisms which are themselves. Consequently a far greater degree of proof and repetitive demonstration is required before any generalisation from scientific biology will be accepted and acted upon by all the people all the time. And so scientific knowledge generally applicable to human eating, learning, loving, mating, ailing, fighting, and so on, has to pass almost the total consensus of opinion in each community before it becomes generally operational. It is therefore unusually important in this field for us to publicise received knowledge in every way we can, particularly to the young. The usual reticence of science and medicine is inappropriate here. I think Dr Boyden agrees with this.

NIGEL ORAM In his very stimulating paper, Dr Boyden has raised a number of problems which we face in the world today. I wonder whether we could discuss the part which we, as academics, should play in trying to find a solution of these problems, beyond the pursuit of our particular research interests.

Dr Boyden has mentioned the accelerated rate of change which is occurring in our world—my sense of urgency may be coloured by my long concern with developing countries, but the pace seems likely to quicken. Who would dare to envisage the state even of our Western societies in, say, ten years' time? Problems of which we are already aware, such as the rapid growth of population in developing countries, will have become more serious: new problems will have arisen. Even now, institutions which many have regarded as the foundation of our social structure—the churches, our form of democratic government, marriage as a legal institution—are under strong attack by students and others who feel that they no longer meet present day needs. With Dr Furnass, I feel that the students' perception of our society, which may be fresher and more acute than my own, should be treated with respect, both because they are the products, and also the agents, of change. In conditions of rapid change, there can be no safety in standing still, and intellectual boldness seems to be needed to confront the social and other problems which threaten us.

Before action, appreciation of the situation: how far are our present methods of research adequate to provide this? Until recently, it might reasonably be assumed that studies within a single discipline, for example sociology, would be sufficient to provide an understanding of the majority of social problems. This is now clearly not so. As Dr Boyden has shown us, for example, biological factors are now seen to be involved in fields which until recently were considered to be exclusively the fields of the social sciences. Ethology may have much to tell us about human behaviour: we have yet to achieve any real understanding of the problem of lawlessness and violence. If, as first suggested to me by Dr Freeman and discussed during the meetings, there is a biological foundation for the persistence of kinship ties, this may provide a significant contribution to the debate on the future of the family.

Failure to study all aspects of a problem may have serious results. From my own experience, continuing deterioration in urban living conditions in developing countries seems to be largely due to conservative attitudes and a failure to study the interrelationship between the major factors determining their growth. In particular, one of the greatest barriers to the improvement of housing conditions has been the insistence of medical authorities on the retention of building and sanitary regulations based on conditions in nineteenth-century Europe. As a result of a lack of financial and other resources, a few people enjoy high-standard living conditions while the remainder live in conditions of chaos. There has been much research into the part which urban environments play in the aetiology of different diseases. There appears to have been little attempt to apply the results of this research to the problem of improving living conditions in areas where resources of all kinds are limited.

What seems to be needed for the understanding of social problems is that:

(1) expertise provided by all relevant disciplines should be brought to bear; and

(2) a synthesis of the different findings should be attempted.

Whether this research should be carried out within universities, or whether mainly through new institutions designed to investigate a particular field, seems to demand serious discussion. No one could deny the need for specialised research in separate disciplines. There is often still, in universities, a desire to defend the territory of a particular discipline against intruders and an intellectual climate which is hostile to interdisciplinary research. While a number of specialist bodies are being established, it is not clear whether they themselves are sufficiently interdisciplinary in character.

Whether academic search for truth is jeopardised by political activity has long been the subject of debate. Many of the problems of the kind raised by Dr Boyden, such as the pollution of the atmosphere and the planning of towns, could almost certainly be solved or ameliorated if there was sufficient public pressure for action to be taken. But a forcibly expressed public opinion on these issues is lacking. For the Greeks, the 'good life' could only be lived in the *polis*, the public realm. For perhaps a large majority of people living in Western countries, their only act as citizens is, unwillingly, to pay their taxes. There are many ways in which an individual can become involved, but all are likely to be concerned with ends and the making of moral judgments. There does seem to be a need for scholars, however, who represent the best-informed sector of the population, to go down into the market place to debate the issues of which they have a particular understanding. Many people, including myself, have little knowledge of the problems which Dr Boyden has discussed and enlightenment can, surely, only be achieved through continuing debate. I am not, of course, suggesting that knowledge itself, as Sir MacFarlane has pointed out in connection with cigarette smoking, is sufficient to provide a solution.

I am therefore offering for discussion the propositions which arise from the very existence of this meeting:

(a) if the environment in which we live is to be improved and indeed if the major problems which we face are to be overcome, there is a need for more problem-oriented and multidisciplinary research; and

(b) there is a need for greater participation by all, but especially by intellectual leaders, in public affairs.

Discussion

HIATT A subject that has not been discussed at this meeting is the role of women in industrial society. In hunter-gatherer communities and in many horticultural and agricultural communities, women not only bore and raised children but also played a vital economic role. Australian Aboriginal women, for instance, provided about two-thirds of the diet by gathering wild vegetables, shellfish and so on, and they continued to do this kind of work until they were quite old. In our society, middle and upper-class women have for a long time been regarded as unfitted for roles in the industrial process and have been encouraged to specialise solely in the reproductive and rearing arts. Lately, for various reasons, the scale of even this operation has been reduced. At the same time, medical technology has increased life expectancy. The upshot is that, whereas girls in pre-industrial societies were trained for a lifetime of motherhood and vital economic services, many girls in our society are groomed only for a relatively short period of motherhood, with little forethought given to how they might occupy the long years between then and the grave. When, on top of this, one thinks about domestic labour-saving devices, it becomes apparent that the average woman today fulfils her destiny in a brief moment of glory and then fades out for the next thirty years or so. This, on the face of it, would seem to be sad, wasteful, and 'bad for the system'.

FREEMAN I would like to make a comment at this stage of a philosophical kind. It bears on what Dr Boyden, Dr Hackett, and Mr Oram have been saying. I think we have reached the stage where there is, at last, the possibility of the emergence of a unified science of man. If we are to bring this off, we shall have to learn to think in some new kinds of ways. We shall have to transcend the limitations of our existing concepts and we shall have to grow out of the idea of finite disciplines as also of the much-vaunted 'multidisciplinary approach'. That great philosopher, Humpty Dumpty, arrived at the profound realisation that words are not the properties of things and, equally, disciplines are not the properties of nature. They are rather of our own contriving, and the artifacts of our ignorance. Now, if we are to attain a unified science of man, we shall have to develop the capacity to consider all of the main determining variables in a single sustained analysis, and it seems to me that this realisation is just beginning to emerge in this present symposium. As one considers the different contributions, the master concept would seem to be adaptation in an evolutionary frame of reference. I think that if we begin to use this as a unifying concept, the possibilities of a unified science of man are probably in time realisable. I am convinced that only when we have attained some such unified science shall we be able to comprehend and to solve the complex problems that face us.

WELLS Mr Chairman, I think that Dr Boyden has raised an issue which has not really been recognised here. Are we mainly concerned with ourselves as individuals or with the future of mankind, because what we do will be different in the two cases. Now if we accept that the past and future evolutionary development of mankind is at least in part a genetic process as well as being partly a cultural adaptation, I suppose we would have to go on from that and accept that within reason, the faster the turnover of generations the quicker and the more effectively this genetic evolution can take place. Now it is true that up to a point this is counterbalanced by the fact that the presence in a population of a proportion of older people may actually assist in cultural adaptation by keeping a core of balanced and experienced knowledge available, but nevertheless in any given set of circumstances there must be some sort of optimal age distribution of the population from the point of view of producing the most rapid and satisfactory evolutionary development. Now if this is so, it would surely seem likely that any species such as man would have developed a mechanism for ridding itself of substantial numbers of old people in the population, particularly after they have had the opportunity of having a few children and producing one generation. I do not know what these mechanisms are and I wonder whether Dr Boyden has any views on this and whether he thinks in particular that arteriosclerosis may be just one of the mechanisms. Because if this is so certain conclusions, I think, follow. Firstly, of course, we should perhaps not strive, if we are worried about the future of mankind rather than about our own lives, too hard to keep alive vast numbers of old people in the population. And secondly, if in fact this is possibly one of the mechanisms that have been developed to assist in rapid and effective evolution, we perhaps should not strive too hard to find remedies for it.

MIMS I would like to make one point about this problem of population control in primitive peoples. Do we know whether primitive people practise contraception other than by abstaining from intercourse during pregnancy, lactation or at other times? We must find out more about the natural regulation of populations. As an illustration of how little we know, some work on environmental temperature and conception suggests that in big cities in the USA[1] and in Hong Kong,[2] there is a fall-off in the rate at which children are conceived when temperatures are above 70°F. Not only this: if you *are* conceived in the warmest months of the year, you are more likely to end up in an institution for the mentally deranged[3] and you are also less likely to end up in the pages of *Who's Who*.

[1] C. A. Mills and F. A. Senior, 'Does Climate Affect Human Conception Rate? *Arch.Int.Med.*, **14**, 1930, 921-9.

[2] K. S. F. Chang *et al.*, 'Climate and Conception Rates in Hong Kong', *Human Biol.*, **35**, 1963, 366-76.

[3] B. Pasamanick and H. Knolbloch, 'Seasonal Variation in the Births of the Mentally Deficient: A Reply', *Am.J.Publ.Hlth*, **50**, 1960, 1737-42.

HETZEL Just two comments, Mr Chairman. First of all, this question of women after the reproductive period. I think there is increasing recognition of this problem and some attempt is being made to mobilise this resource in man-power in this whole field of personal work with people in the new communities of the modern urban world. In the United States under the Kennedy administration certain experiments were initiated in training women of this age as mental health assistants in the whole program of prevention and maintenance of mental health. And I believe that this is one important avenue where women can be absorbed and find a new life, so to speak, in the urban community.

I would like to support what Dr Freeman says, although I realise there are always splitters and synthesisers in any gathering and they never really can communicate. I am very much with him; I think this is the orientation we have lost in medicine to some extent and we need it. However, this is a highly controversial statement.

HIPSLEY I would like to ask Dr Boyden to comment on a problem that concerns me profoundly and this is: Do these corrective adaptations to maladjustment ever conflict? It seems to me that you can sit down on a problem-orientated basis and work out a perfect solution for each problem but when you try to work out solutions for the lot you find that it is just not on. For example, the control of infectious diseases I think has been brought about by corrective adaptations, but this has been the main factor responsible for the increase in population and the threatening food shortage. Now we are led to believe today that contraception may be the answer to this problem but I strongly suspect that if it is applied on a large scale, it too will have repercussions in the health field and require further action. I think it was actually Professor Dubos who made the remark in his book[1] that there is no paradise on earth for men and that to eliminate maladjustment would be to eliminate adventure, creative thinking and to elect for boredom and stagnation.

[1] R. Dubos, *Mirage of Health*, New York, 1959, Harper and Brothers, 235.

BOYDEN Firstly, I am very pleased to see that, for all his irrationality, Rousseau's name has been mentioned a number of times in this symposium. I am always surprised by the vehemence with which his ideas are often condemned. I find this interesting in itself, but it is a subject which I will not attempt to develop at this point of time. I will simply say that I am not as strongly anti-Rousseau as some. The Hobbes-Rousseau pendulum swings back and forth, and no doubt the truth lies somewhere in between the two extremes.

Secondly, I would like to say I am in complete agreement with Dr Hackett's emphasis on the genetic diversity of the human species. In the responses of the human organism to various changes in environmental conditions, we see many examples of differences both in the degree of response to given environmental changes and in the type of response in different human beings—due partly, no doubt, to previous experience (that is, to environmental influences) but also certainly due in part to genetic factors. For example, take the introduction of refined carbohydrates into the diet. It is fairly well recognised that there are genetically-determined differences with regard to the responses to this environmental change. For example, some genotypes are more likely than others to develop dental caries. Similarly, some genotypes respond to the introduction of refined carbohydrates by developing diabetes, a disease which, while clearly controlled by a genetic factor, does not apparently occur in populations which do not have refined carbohydrates in their diet.

I do not agree, of course, with Dr Hackett's statement that when the public health movement got under way in the last century, there was '*no* understanding' of the nature of disease—in our biological sense. There certainly was a degree of ecological understanding, in that it had been noted that the disease states were correlated with certain characteristic environmental conditions.

I agree wholeheartedly with the views expressed by Mr Oram and Dr Freeman emphasising the urgent need for a new approach to the study of man. On the one hand we have the processes of nature which are studied by natural scientists, and on the other hand we have cultural processes and cultural phenomena which are studied by social scientists and students of the humanities. This dichotomy is justifiable to an extent because these two sets of phenomena are distinct in many ways. Nevertheless, there is a tremendous and extremely important interaction between them, and it is this interaction which we have been interested in today, and which I have been stressing in my paper.

Dr Wells suggests that arteriosclerosis is in fact a biological mechanism selected for in human evolution because of its species survival value in removing part of the population after the reproductive period is over. I think this is a hypothesis worth bearing in mind in our attempts to understand the biology of man. My personal view at the present point of time is that the hypothesis is probably wrong; but that if further information in its favour comes to light I might change my mind. I can see no biological advantage in the elimination of older individuals in a population unless they are a burden on the population as a whole as a *result* of their incapacitation through arteriosclerosis, high blood pressure and so on. Such incapacitation *per se* (even if it did lead to early death), could not be of survival value. In other words, the optimum age distribution depends on the state of health of the various age groups, and not vice versa. I wonder if Dr Wells would say the same, for example, of lung cancer? Some of these pathological conditions are so obviously consequences of conditions of civilisation. I agree that it has yet to be conclusively proven that cardiovascular disease is a disease of civilisation—although I belong to the school that considers the weight of evidence is strongly in favour of this possibility. But if it could be shown that it is as common a condition in all human populations, whatever their conditions of life and including relatively 'uncivilised' groups, as in modern urban society, then this would be a good argument in favour of Dr Wells's suggestion. However, there is good evidence that this is not the case.

Dr Hipsley asked whether I would agree with him that forms of corrective adaptation tend sometimes to conflict with each other; I would certainly agree that under existing conditions this may happen, although in general they are less likely to have undesirable repercussions than antidotal forms of adaptation. But, as I mentioned, corrective adaptive processes are not always feasible for social or economic reasons and even when they are feasible they might have some undesirable repercussions under the new conditions.

I think there is the suggestion in Dr Hipsley's remarks that we in fact need a certain degree of maladjustment in society in order to make life worth living. I would agree that a certain level of frustration would seem to be a desirable thing. On the other hand I also think that a great deal more effort should be put into attempts to understand the human situation in scientific terms, and this must involve learning more about the processes and mechanisms of cultural adaptation to biological maladjustment—and also, of course, about the causes of these forms of maladjustment. I do not think we need worry for a long while about a situation arising in which we will not have any forms of biological maladjustment.

9 René Dubos

The Impact of Civilisation on the Biology of Man

The Biology of Civilisation—with Emphasis on Perinatal Influences

The word civilisation is of recent origin; it probably appeared in print for the first time in 1757. As used throughout the eighteenth century, it meant better ways of life, humane laws, limitations on war, a high level of conduct and purpose. As late as 1772, Samuel Johnson refused to accept the word in his dictionary because he felt that the older word civility had the same meaning.

I have introduced this bit of pedantry at the beginning of my presentation to illustrate that civilisation involves not only the various technologies that are presently used to manipulate the external world, but also the social practices and cultural traditions that govern human life. The broader meaning of the word civilisation is relevant to our symposium, because man is affected by his attitudes, beliefs, and hopes, as much as by the forces that impinge on him from the outside.

Civilisation includes all the processes through which human life has become different from animal life—a differentiation that was already well advanced in the Old Stone Age, long before the neolithic revolution. In fact, there cannot be any *human* life without civilisation. Humanness and civilisation are coeval. The interplay between the purely biological attributes of *Homo sapiens* and his total environment accounts for both the desirable and the objectionable manifestations of humanness.

The philosophers of the Enlightenment would have been surprised by the somewhat pessimistic tone of our symposium, emphasising as it does the dangers of civilised life. They believed that an era of greater civilisation, and therefore of better human life, was at hand because the advances in science and technology that were occurring in their times made it appear that man would soon be able to control his own nature as well as external nature, and thus to choose his destiny. It is obvious that something has gone wrong with this hope during the past 100 years. Increased control of the environment and economic prosperity do not necessarily result in better health or greater happiness. The modern ways of life create new problems of their own.

The papers presented at this symposium reflect the efforts made by the scientific community to understand why civilisation—which was supposed to improve human life—in fact creates a social and technological environment that is often deleterious to man. I believe that all the participants tacitly accept the following explanation for the partial failure of the new world created by science and technology to provide biological conditions optimum for human life.

The structure of man's body and brain has not changed significantly since the late Stone Age. By that time, *Homo sapiens* had been shaped by the environmental forces that had prevailed during his evolutionary development; it is certain furthermore that his genetic endowment will remain essentially the same during the foreseeable future. But while modern man cannot change genetically, he has to live now under the conditions created by social and technological forces; these conditions differ profoundly from those under which he evolved and to which he had become genetically adapted.

Fortunately, the genetic properties that have developed in the various components of the human gene pool during the evolutionary past are extremely numerous; their redistribution and selection can probably result in a fairly rapid genetic drift whenever the selective pressure is strong enough. But granted that man's range of phenotypic variability is extremely wide, it is limited nevertheless by the biological characteristics encoded in his genetic equipment. In practice, many social and technological innovations impose on him adaptive demands that are often excessive—hence the diseases of civilisation.

Health and disease are the expressions of the relative degrees of success or failure experienced by man as he tries to respond adaptively to environmental challenges, and also to the inner demands created in him by traditions and aspirations. Changes in the ways of life and in the environment now occur so rapidly that genetic adaptation is in practice non-existent in human life, or of trivial importance. But, although man can supplement his biological mechanisms of phenotypic adaptation with even more effective social mechanisms, his adaptive responses often fail—for two different kinds of reasons.

Since his surroundings and ways of life are forever changing, man can hardly ever achieve a state of adaptedness to his environment. The endless and futile effort to achieve such a state commonly involves struggles that result in suffering, disease, and unhappiness. Furthermore, responses that appear to constitute successful adaptations at the time they occur may have secondary effects that are deleterious in the long run.

The phrase 'adaptive response' was coined to deal with phenomena of general biology, but it requires many qualifications when applied to the study of human problems.

The general biologist usually defines the word 'adaptation' in Darwinian terms. For him, the word implies a state of fitness to a given environment, enabling the species to multiply and to invade new territories. According to this definition, man is remarkably adapted to life in highly urbanised and industrialised societies, as shown by the fact that his populations continuously increase and that he spreads urbanisation and industrialisation to more and more of the earth. It is obvious, however, that further population increase has become objectionable, and may soon become intolerable. In applying the concept of adaptation to the human species, it is therefore necessary to use criteria different from those used in Darwinian population theory.

Physiologists and psychologists give to the word 'adaptation' a meaning different from that given by general biologists but their interpretation also fails to take into account the peculiarities of human life.

For physiologists and psychologists, a response is adaptive when it enables the person to maintain homeostasis through metabolic, hormonal, or mental processes that tend to correct the disturbing effects that environmental forces exert on the body and the mind. Such adaptive responses contribute to the welfare of the organism at the time they occur, but they often have secondary effects that are deleterious at a later date. Many, if not most, chronic disorders are the secondary and delayed consequences of adaptive responses that were useful at first, but are faulty in the long run. When evaluated over man's whole life span, homeostatic mechanisms are therefore less successful than commonly assumed.

It has long been recognised of course that homeostatic mechanisms can lead to unhomeostatic effects. This is particularly true when the homeostatic response is excessive. In traumatic shock, for example, intense vasoconstriction is homeostatic to the extent that it preserves blood pressure, but it is unhomeostatic eventually because it deprives the kidney and other essential organs and tissues of their vital blood flow. In the hypervolemia of heart failure, the congestive state is useful up to a point in filling a weakened heart chamber, but leads ultimately to total failure by way of both excessive vascular pressures and overdilated heart chambers. As to the inflammatory reaction, it may help in fixing or destroying the aggressive agent, but it can destroy the organ while attempting to protect the body.

As already mentioned, homeostatic mechanisms commonly have delayed and indirect consequences responsible for the pathology of many chronic disorders. The production of scar tissue is a homeostatic response because it heals wounds and helps in checking the spread of infection. But fibrosis in the liver or in the kidney means cirrhosis or glomerular nephritis; scar tissue may freeze the joints in rheumatoid arthritis or may choke the breathing process in the lung.

Many other examples readily suggest themselves, such as the various forms of hyperimmune response and the so-called compensatory reactions such as compensating polycythemia or compensating emphysema. These processes exert a protective or reparative function when they first occur, but they can become destructive in the long run. All too often the wisdom of the body is a very short-sighted wisdom. The renal pathologist, J. Oliver, stated this problem clearly in his discussion of acute tubular necrosis of the kidney: 'The majority of the fatalities, in fact, occur during the "recovery period", as a result not so much of the processes of renal damage as of the mechanism of renal repair'. The end results of social adaptation are often as destructive as those of biological adaptation.

Science enables technology to do almost anything, but there is a painful discrepancy between what man aims for and what he gets. He sprays pesticides to get rid of mosquitoes and weeds, but he thereby kills birds, fishes, and flowering trees. He drives long distances to recapture the purity of nature, but he poisons the air along the way. He eliminates food shortages through scientific agriculture, but creates thereby new patterns of diseases caused by overnutrition. He synthesises drugs to treat many kinds of physical or mental disorders, but finds that new illnesses commonly result from the use of these very drugs. He builds machines to escape from physical work, but the more efficient the machines, the more exacting the constraints they impose on his life.

The environmental forces—chiefly of technological origin—that now impinge on man's life are so complex and powerful that they appear beyond human control. Many are the persons who believe that we shall have to change completely our ways of life, if we do not want the physical and mental worlds to continue going from bad to worse. New ways of life, however, are difficult to imagine; we no longer believe in utopias. Yet there remains much reason for hope, and even much room for utopian thinking.

Hope is justified by the fact that man is immensely adaptable. We worry about the problems of world nutrition, but should find encouragement in the fact that some people thrive on an essentially vegetarian diet, while others relish an animal diet. We are concerned with crowding, but should not forget that the inhabitants of some of the most crowded countries—such as Holland—or cities—such as Hong Kong—enjoy a high level of physical and mental health. Men have created exciting civilisations in humid equatorial lowlands, and on semi-desertic highlands. Human beings are so adaptable indeed that they can function and multiply even under horrible conditions. The most difficult problem may not be to assure the survival of the human species, but rather to decide what kinds of adaptation are compatible with the maintenance of desirable human values.

Another reason for hope is that the public is now intensely concerned with the dangers created by technology and urbanisation, and is beginning to realise that many environmental effects are indirect and delayed. The bio-medical profession, unfortunately, has not yet focused its attention on this problem. So far, the knowledge of man's responses to his environment has been chiefly derived from the study of a few dramatic accidents, such as caused by ionising radiations, drug intoxication, or environmental pollution. Research in environmental biology has been stimulated also by the need to train combat forces for operation in the tropics or the Antarctic, to overcome the new kinds of fatigue and boredom caused by automation, or to manage the effects of sensory deprivation experienced by the crews of submarines or space vehicles. The study of these and other specialised problems has provided useful information, but of an episodic character. Its chief merit has been to make clear that the new human problems emerging in technological societies cannot be dealt with by applying existing knowledge derived from atomistic biology but will require a new scientific approach.

During the past few decades, the greatest biomedical advances have been achieved through the detailed analysis of isolated cellular and biochemical systems. Valuable as it is, the information thus obtained is not sufficient to recognise, let alone understand, man's responses to environmental stimuli and insults. Any environmental problem implies situations in which several interrelated systems operate simultaneously. Of special importance in this regard is the fact that the response to any stimulus is conditioned by the past experience of the organism. The experiential past, in other words, is as important a factor in biological response as is the genetic endowment, the structure of tissues, and the nature of the stimulus. Multifactorial investigations of man's responses to his environment will demand new conceptual and experimental methods, very different from those involving only one variable which have been the stock-in-trade of experimental science.

The advent of computer technology may help in analysing epidemiological observations, to learn more of human responses to the total environment. It is certain, furthermore, that environmental biology can become an experimental science because animal life provides models presenting a close analogy to almost any problem of human life. This is true not only for anatomical characteristics, physiological attributes, pathological states, behavioural patterns and social organisation but also for the conditioning of adaptive responses by past experiences. The following list indicates a few areas of environmental biology that lend themselves to experimental investigation through the use of animal models:

1. Indirect and delayed effects of biologically active substances, such as drugs and environmental pollutants.
2. The effect of sensory stimuli—or lack of—and of other environmental conditions, on the development of sense organs and on physiological processes.
3. The effects of relative degrees of isolation or crowding on hormonal activities and behavioural patterns.
4. The acquisition of tolerance to injurious agents and situations; its distant consequences.
5. Enduring effects of perinatal influences, i.e. of the factors (nutritional, infectious, toxic, behavioural, etc.) that impinge on the organism during critical periods of its development (pre-natal and early post-natal).
6. The conditioning of the response to any particular insult (of physiochemical, infectious, or behavioural origin) by the effects of the total environment on the host; i.e. the role of non-specific etiological factors in epidemiology.
7. The need to keep social and technological innovations within a framework determined by the biological limitations of man.
8. The adaptive potentialities of the human organism at various stages of its development.

The phrase 'adaptive potentialities' has been introduced above to convey the view that the environmental study of human life should not be limited to a consideration of pathological phenomena. Each person has a wide range of physical and mental potentialities that remain untapped. These become expressed phenotypically only to the extent that the person is given a chance to respond adaptively to the proper stimuli under the proper circumstances. As we shall now see, the responses to environmental stimuli made by the organism during the early phases of its development are of special importance, because they affect profoundly and lastingly the anatomical, physiological, and behavioural characteristics of the adult.

In his essay on the 'Uses of Great Men', Ralph Waldo Emerson wrote a statement that has a direct bearing on the shaping of human personality by the total environment: 'There are vices and follies incident to whole populations and ages. Men resemble their contemporaries *even more* than their progenitors' (italics mine).

As a moralist, Emerson was primarily concerned with the intellectual and moral attributes of human beings; but his aphorism is just as valid for physical and physiological attributes. We resemble our progenitors because we derive from them our genetic endowment. Genes, however, do not really determine the traits by which we know a person; they only govern the responses that the person makes to environmental stimuli. Individuality progressively emerges from these phenotypic responses.

Whereas the genetic pool of a population remains essentially constant, the environment changes rapidly. We resemble our contemporaries because the phenotypic expressions of the genetic endowment are determined by the environmental forces that impinge more or less simultaneously and with almost equal strength on all the members of a given generation in a given social milieu.

The phenotype is constantly being moulded by the environment throughout the whole life span. Individuality could be defined as the continuously evolving phenotype. Early influences, however, certainly play the most important role in converting genetic potentialities into phenotypic reality.

As commonly used, the phrase 'early influences' denotes the conditioning of behaviour by the experiences of early life. But early experiences do more than condition behavioural patterns; they also affect, profoundly and lastingly, other biological characteristics such as initial growth rate, efficiency in the utilisation of food, anatomic structures, physiologic attributes, maximum adult size, resistance to infection, response to various forms of stress and stimuli, in brief, almost every phenotypic expression of the adult.

A few examples from contemporary life will suffice to illustrate the effects of early influences on human populations.

Japanese teenagers are now much taller than their parents and differ in behaviour from their pre-war counterparts, not as a result of genetic changes, but because the post-war environment in Japan is very different from what it was in the past. A similar phenomenon is observed in the settlements of Israeli kibbutzim. The kibbutzim children are given a diet and raised under conditions as favourable as can be devised; within one generation, they tower over their parents who originated from the crowded and unsanitary ghettos of Central and Eastern Europe.

The acceleration of growth in Japan and in the Israeli kibbutzim constitute but particular cases of a constant trend toward earlier maturation of children in Westernised countries. This is evidenced by greater weights and heights of children at each year of life, and by the earlier age of the first menstrual period. In Norway, for example, the mean age of menarche has fallen from 17 years in 1850 to 13 in 1960; similar findings have been reported from the USA, Sweden, Great Britain, and other affluent countries.

Growth is not only being accelerated; the final adult heights and weights are greater as well as being attained earlier. Some fifty years ago, maximum stature was not being reached in general until the age of 29; commonly now it is reached about 19 in boys and 17 in girls. With regard to the age of puberty, the change seems to consist in the restoration of the developmental timing that prevailed several centuries ago and that had been greatly retarded by the ways of life at the beginning of the industrial revolution.

The factors responsible for these dramatic changes in the rate of physical and sexual maturation are not completely understood. There are good reasons to believe, however, that the control of childhood infections and improvements in nutrition of the mother and of the child, have played a large part in the acceleration of development. This change in turn has probably been responsible for the larger size achieved by adults.

I shall now briefly describe experimental models that illustrate the effects of early influences on the development of laboratory animals.

Pioneering investigations on the lasting biological effects of early post-natal influences were carried out in England during the 1960s by R. S. McCance and Elsie M. Widdowson of Cambridge University. These investigators compared the growth of rats suckled in small litters (3 young per lactating female) with that of comparable animals suckled in large litters (18 young per female). They found that the rats of the latter group became much smaller adults than those of the former group, even though the animals of both groups were given unlimited food after weaning. It can be assumed that the animals raised in small litters developed more rapidly than the others because they enjoyed a nutritional advantage during the lactation period. As McCance and Widdowson pointed out, however, the interpretation of their findings may be more complex than appears at first sight. The design of their experiments did not rule out the possibility that infectious and psychological disturbances occurred when lactating females had to nurse large numbers of young. Such disturbances might have played a role in the retardation of growth observed in this group.

In our own studies, an attempt was made to eliminate or minimise the disturbing effect of indirect, non-nutritional factors. To this end, the experiments were conducted with specific-pathogen-free (SPF) mice so as to avoid the activation of latent pathogens by nutritional deficiencies. In all cases, the litter size was reduced to 8 young, which is physiologically normal for mice. Whenever possible the animals were put on the experimental diets two weeks before mating so as to eliminate the behavioural disturbances that commonly occur at the time of a dietary change.

The principle of the experiments consisted in introducing the conditioning factor (infection, nutritional deficiency, toxic influence, or behavioural disturbance) early in the life of the animal—either during gestation or lactation. Immediately after weaning, all young animals were placed under exactly the same optimum conditions of husbandry and kept undisturbed during their whole life span. This made it possible to recognise the lasting effects of the early manipulation on characteristics such as rate of growth, maximum adult size, longevity, and response to stimuli.

SPF mice weigh more at weaning time than do mice of the same genetic origin produced under less sanitary conditions. Furthermore, they reach larger adult size and exhibit more desirable physiological characteristics as well as greater resistance to various forms of stress. These differences are not due to genetic changes as is shown by the following experiments.

In one type of experiment, new-born SPF mice were contaminated orally one or two days after birth with material obtained from ordinary mice that were healthy according to usual criteria but had been raised on commercial farms. The contamination did not cause any obvious sign of disease, not even diarrhoea. Yet the contaminated animals were smaller than the uncontaminated controls at weaning time, and grew into smaller adults.

The following facts strongly suggest that this growth-depressing effect is associated with viral rather than bacterial multiplication in the intestinal tract. When SPF mice are contaminated orally shortly after birth with certain bacterial cultures isolated from the intestinal contents of adult ordinary mice, these bacteria multiply extensively throughout the gastro-intestinal tract and persist at extremely high levels until weaning time. However, such bacterial infections do not affect significantly either weaning weight, growth rate, or maximum adult weight.

In contrast, weight depression can be consistently brought about by contaminating new-born SPF mice with a bacteria-free filtrable agent isolated from the intestines of ordinary mice.

Only very young animals (preferably less than 3 days old) prove susceptible to the weight-depressing effect of this agent, but the effect is transmissible from one generation to the next.

A profound and lasting depression of growth can be produced by administering bacterial endotoxin to the mother of SPF mice, either during gestation or during lactation; this effect appears irreversible. Although marked weight loss can also be produced by giving endotoxin to young animals after weaning, the depression of growth thus achieved is rapidly and completely reversible. This finding is compatible with the view that environmental factors are more likely to alter biological characteristics in an enduring manner when they impinge on the organism during certain critical stages of its early development.

A lasting effect on growth can be produced also in the absence of contamination or intoxication by feeding the mouse dam a restricted or deficient diet, either during gestation, or after the birth of her young during the lactation period. Lasting depression of growth has been achieved, for example, by lowering the content of the dam's diet in magnesium, or in lysine and threonine. The growth-depressing effect so achieved persists throughout the whole life span of the young, even when the latter are given unlimited amounts of an optimum diet at weaning time and constantly thereafter. These findings are relevant to human situations since the diets of most underprivileged people consist chiefly of plant products, which are commonly deficient in certain amino acids.

Depression of growth resulting from lysine and threonine deficiency during gestation or lactation does not seem to affect adversely the health of the young, or to decrease their longevity. In fact, the results of two experiments in which the animals nursed by mothers on different diets were kept undisturbed and on optimum diets throughout their whole life span, suggest that the smaller animals had a greater life expectancy than the larger ones.

Even when maintained under optimum conditions, SPF mice exhibit marked individual differences in weaning and adult size. However, the young of each particular mother are usually remarkably uniform. The results of several kinds of experiments, based on foster-mothering, have established that the differences from litter to litter, and the uniformity within each litter, are not determined by the genetic endowment of the young, but rather by some behavioural characteristics of the mother, such as the quality of nest building and attention to the young. Whatever the precise characteristics of 'mothering' quality, they exert on the young effects that persist throughout their life span.

The mechanisms through which early influences can exert such profound and lasting effects are both physiological and mental in nature.

Adult animals exhibiting growth depression as a result of perinatal influences eat as much food when removed from their mothers as do controls of the same age. However, they are less efficient than the controls in utilising the food they eat; their ability to incorporate amino acids is much depressed. Such anabolic deficiencies persist long after the initial disturbance, at least for six months after weaning and perhaps for the whole life span.

The synthesis of protein and of DNA in the muscles and in the brain is also decreased by the early influences that are capable of depressing adult growth. Furthermore, the production of certain types of cells seems to be depressed at a critical stage in development, and this depression cannot be corrected later in life—at least by changes in nutrition or husbandry. For example, adult animals that have been nursed by mothers fed a restricted or deficient diet, or have been contaminated shortly after birth, have fewer lipogenic cells than expected —a fact which may explain why none of them becomes obese. The numbers of muscle and brain cells are also depressed.

The development of the brain is influenced of course by the metabolic factors that govern anatomic and physiologic growth, but also to a surprising extent by the nature and intensity of the stimuli to which the organism is exposed early in life. A proper degree and kind of stimulation accelerates neural and behavioural development, as measured by chemical and enzymatic characteristics of the brain, and by exploratory behaviour and learning ability. In contrast, an impoverished environment results in biochemical and behavioural deficiencies.

The 'imprinting' of birds is almost a caricature of the effect of early influences on behavioural patterns; the techniques of animal training make practical use of the possibility to shape neural and behavioural organisation during early life.

In man, as in animals, physical and mental developments exhibit critical periods, but their timing and duration have not been precisely defined. In any case, since the human body and brain are incompletely developed at the time of birth, there is no doubt that the child completes his development—anatomic, physiologic, and mental—in the course of responding to environmental stimuli. A true science of *biological Freudianism* would include all aspects of biological and mental conditioning by the environmental factors that impinge on the human organism during the critical stages of development.

In the light of these facts, the impact of civilisation on human biology must be considered not only with regard to disease causation but even more perhaps from the point of view of the *formative* effects that civilised life exerts on the human organism. Environmental factors not only affect well-being at the time they are experienced; more importantly they condition future responses to almost any stimulus. They shape the organism physically and mentally by converting genetic potentialities into phenotypic realities.

Although I have focused my remarks on the shaping of physical and mental attributes by environmental forces, this does not imply that I regard man as the product of a kind of predestination. Rather, I wanted to emphasise that choosing or creating an environment is an act that not only affects us in the here and now, but also imposes a pattern on the future. The concept of an optimum environment is unrealistic because it implies a static view of man. Value judgments largely determine where we want to go and what we want to become; in consequence, they are as important as objective, measurable parameters in evaluating the quality of the environment.

That biological criteria must be supplemented with value judgments can be illustrated by the kind of 'undesirable' changes in human life that are likely to occur if man becomes biologically adapted to extremely high population densities. The complexity of social structures will then make some form of regimentation unavoidable; freedom and privacy may come to constitute antisocial luxuries, and their attainment will involve real hardships. In consequence, the human beings most likely to prosper will be those willing to accept a regimented life in a teeming and polluted world from which all wilderness and fantasy will have disappeared. The domesticated farm animals and the laboratory rodent on a controlled nutritional regimen in a controlled environment will then become true models for the study of man.

Human beings are so adaptable that they can survive, function, and multiply despite malnutrition, environmental pollution, excessive sensory stimuli, ugliness and boredom, high population density and its attendant regimentation. But while biological adaptability is an asset for the survival of *Homo sapiens* considered as a biological species, it can undermine the attributes that make human life different from animal life. From the human point of view, the success of adaptation must be judged in terms of values peculiar to man.

The need to consider values in evaluating the impact of civilisation on human values makes utopian thinking almost inevitable. In fact, the formulation of utopias may help to define some of the biological factors that need be incorporated in human values. I shall illustrate this statement by comparing the complementary roles of uniformity and of diversity in the development of civilised life.

No population—animal or human—can survive without some form of hierarchy that integrates its individual components into a well-defined social structure. Man cannot exist as an isolated individual; he must function as member of a group, and therefore accept its values and its goals. Societies cannot remain viable without a large measure of uniformity among their individual members and in their environment.

On the other hand, since human potentialities can be realised only to the extent that circumstances favour their phenotypic expression, it is desirable for the progressive unfolding of civilisation that the social environment be as diversified as safely possible.

If the surroundings and ways of life are highly stereotyped—whether in prosperity or in poverty—the only components of man's nature that become expressed are those adapted to the narrow range of prevailing conditions. Hence the dangers of many modern housing developments which, although sanitary and efficient, are designed as if their only function was to provide disposable cubicles for dispensable people.

Uniformity is thus essential for the safe functioning of the group, and diversity for the more complete development of the individual. Uniformity and diversity constitute two equally essential components of true functionalism. Both must be considered in the formulation of social values because civilised societies commonly suffer from a lack of balance between these two fundamental attributes of human life.

Science cannot define or impose values; but it can go far towards predicting the likely biological consequences of behavioural, social, and technological practices. It can thereby contribute to the formulation and evolution of values by warning man of the consequences likely to result from a certain course of events, and thus providing him with more factual basis for options.

Index

Designed by the ANU Design Section
Text set in 10 point Baskerville one point leaded
and printed on 85 gsm Burnie English Finish
by Halstead Press Pty Ltd, Sydney